MINI

ENCYCLOPEDIA

BUGS

ENCYCLOPEDIA

BUGS

Miles
Kelly

First published as *Bugs* in 2009 by Miles Kelly Publishing Ltd
Harding's Barn, Bardfield End Green, Thaxted, Essex, CM6 3PX, UK

Copyright © Miles Kelly Publishing Ltd 2009

This edition printed 05/14

LOT#:
2 4 6 8 10 9 7 5 3 1

Publishing Director Belinda Gallagher
Creative Director Jo Cowan
Series Designer Helen Bracey
Cover Designer Jo Cowan
Volume Designers Stephan Davis, Rocket Design Ltd
Picture Researchers Jennifer Cozens, Ned Miles
Indexer Eleanor Holme
Production Manager Elizabeth Collins
Reprographics Stephan Davis, Anthony Cambray, Jennifer Cozens
Assets Lorraine King
Consultant Clint Twist

ISBN 978-1-4351-5640-1

Printed in China

British Library Cataloging-in-Publication Data
A catalog record for this book is available from the British Library

Made with paper from a sustainable forest

www.mileskelly.net
info@mileskelly.net

Contents

Insect world

Ants, bees, and wasps

Butterflies and moths

Flies, beetles, and bugs

Termites, cockroaches, crickets, and grasshoppers

Other insects

Spiders, scorpions, and mites

Worms, millipedes, and centipedes

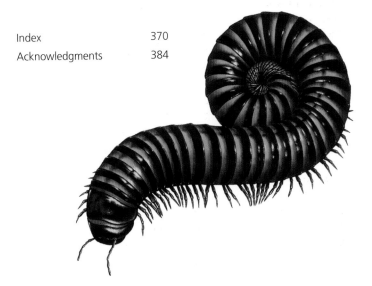

Insect world

Classification

Experts have divided all living things on earth into different groups and given them special names. Each group shares certain features.

All living creatures are classified into five major kingdoms—animals, plants, fungi, monerans (such as bacteria), and protists.

▼ *The phylum Arthropoda includes insects, spiders, and centipedes. Insects were the first creatures on earth to fly, more than 350 million years ago (mya).*

1 Spiny green nymph	**4** Dragonfly	**7** Pink-winged stick insect	**10** Giant long-horn beetle	**13** Tarantula hawk wasp
2 Tiger swallowtail butterfly	**5** Cricket	**8** Tarantula	**11** Honeybee	
3 Cockchafer beetle	**6** Stag beetle	**9** Millipede	**12** Wood ants	

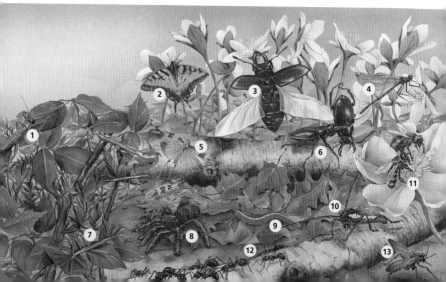

▶ *Trapped in sticky tree resin millions of years ago, this delicate insect was beautifully preserved as the soft resin turned into hard amber.*

These kingdoms are further divided into phyla (singular: phylum). There are around 20 phyla in the animal kingdom. The phylum Arthropoda is the largest in the animal kingdom and includes insects, spiders, crabs, and centipedes.

The different phyla are split into smaller groups called classes. Birds, mammals, reptiles, and insects all belong to different classes. Insects belong to the class Insecta. Spiders and scorpions belong to the class Arachnida.

There are different orders within each class. For example, ants, bees, and wasps belong to the order Hymenoptera within the class Insecta. All hymenopterans have wings and can sting.

The study of insects is known as entomology. Entomologists have divided all the insects discovered so far into 32 orders.

Orders are further subdivided into families. Lions, tigers, and pet cats belong to the family Felidae. Similarly, all ants belong to the family Formicidae.

Families are even further divided into genus and species. Only animals that belong to the same species can mate together to reproduce fertile offspring.

Scientists called taxonomists give scientific names to living things. These names are in Latin and include the genus and species name.

Insect fact file

🐝 **There are between** one and eight million species of insects, compared to just one human species.

🐝 **Insects are successful survivors** for different reasons, such as their powerful exoskeletons, their ability to fly, and their small size.

🐝 **Some insects** can fly for long distances. Certain butterflies migrate thousands of miles to avoid bad weather.

🐝 **Cockroaches** have been living on earth for around 300 million years. Today's cockroaches look very similar to those living hundreds of millions of years ago.

🐝 **Insects** serve as the largest source of food for other animals.

🐝 **Fairyflies** are the smallest insects in the world. They are only 0.008 inch long—that is the size of the period at the end of this sentence.

🐝 **People have domesticated** silkworms for so long that these insects do not exist in the wild anymore.

🐝 **Insects are cold-blooded** animals, so their growth and development depends on how hot or cold the weather is.

🐝 **Scientists have developed** "insect robots" that copy the agility of real insects. These robots are used to explore dangerous areas, such as minefields and the surface of other planets. Robots are not nearly as agile as real insects, but mimic the way they move.

DID YOU KNOW?

Scientists discovered the fossils of a dragonfly that lived 300 mya. It had a wingspan as big as a seagull.

▼ Together, moths and butterflies, such as this swallowtail butterfly, make up one of the largest groups of insects. There are ten times more moths than butterflies.

Record-breaking insects

Mayflies have the shortest lives. As nymphs, they spend two to three years at the bottom of lakes and streams. When they emerge on dry land as adults, they live for an average of one or two days.

Locusts are the most destructive insects. The desert locust is the most damaging of them all. Although they are only about 2 inches long, they can eat their body weight in food every day. One ton of locusts, a fraction of a swarm, can eat the same amount of food in one day as around 2,500 people.

The calling song of the African cicada measures at 106.7 decibels (as loud as a lawn mower), making it the loudest insect. The male cicada produces loud buzzing sounds by vibrating drumlike membranes on its abdomen.

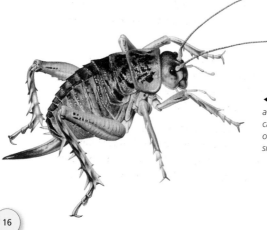

◀ The giant weta, a type of cricket, can weigh up to 2.5 ounces—as much as a small bird!

Female Malaysian stick insects lay eggs measuring 0.5 inch—larger than a peanut! Insects such as cockroaches lay egg cases that are larger than this, but they contain about 200 individual eggs.

The highest insect jump measured was 27.5 inches by the spittlebug. When it jumps, the spittlebug accelerates at 8,900 mph and overcomes a g-force of more than 414 times its own body weight.

The world's largest known spider is a male Goliath bird-eating spider, collected by members of the Pablo San Martin Expedition at Rio Cavro, Venezuela, in April 1965. It had a leg span of 11 inches—as big as a dinner plate!

With a wingspan of 12 inches, the atlas moth is the biggest recorded moth. It can often be mistaken for a bird.

The mother-of-pearl moth caterpillar can travel at 16 inches per second. It is the only creature that deliberately rolls away when attacked.

The larva of the North American Polyphemus moth eats an amount equal to 86,000 times its own birth weight in the first 56 days of its life. For a human, this would be equivalent to an average baby eating 273 tons of food.

Measuring up to 18 inches, the large "human" roundworm infects about one billion people worldwide. Humans can be infected with as many as 20 worms. A female worm produces about 300,000 eggs each day and about eight million in its lifetime.

Insects and people

Insects were always of great importance to human civilizations. People rear insects such as silkworms and honeybees to obtain important materials from them, such as silk, honey, and wax.

Archaeologists have discovered prehistoric cave paintings that show scenes of honey collection and the extraction of honey from beehives.

Japanese samurai warriors painted intricate butterfly patterns on weapons and flags to symbolize nobility.

Cicadas and crickets have been captured and reared by humans for the beautiful sound they can produce. The ancient Chinese regarded cicadas as a symbol of immortality and rebirth.

Many insects, such as mosquitoes, lice, and bedbugs, feed on human blood.

Some people eat insects such as termites, cicadas, leaf-cutter ants, and water bugs, and consider them to be delicacies.

Deadly diseases such as the bubonic plague were transmitted to humans by insects such as fleas, and caused millions of deaths.

Doctors used to insert maggots in wounds to eat dead flesh and disinfect the wounds by killing bacteria.

Some insects are pests and can cause serious damage by destroying crops and fields. However, many insects feed on agricultural pests and help farmers.

▶ People began to domesticate bees about 3,000 years ago. Today, honeybees are kept in beehives containing removable frames. The bees store honey in the upper frames, which the beekeeper removes to harvest the honey.

Forensic entomology

🪰 **Insects can help detectives** and other experts to solve crimes. There is a special science in which scientists study the insects found on dead bodies. This is called forensic entomology.

🪰 **A lot of changes** take place in the body once a person dies. The corpse becomes host to a variety of insects.

🪰 **Blowflies** are the very first insects that normally infest a corpse. They lay eggs that hatch into maggots. Other insects soon find their way to the corpse, in search of these maggots and flies.

🪰 **Forensic experts** can guess the time of death by studying these insects. They study the life cycle of the flies and determine the time of death. For instance, if an expert finds maggots on the corpse, they can be sure that death occurred a few days previously. This would give the flies enough time to lay eggs and then the eggs to hatch into maggots.

🪰 **The cause of death** can also be detected by studying the insects on the corpse. Sometimes, experts spot maggots and eggs on cuts and wounds in the skin that may have been fatal.

🪰 **Sometimes**, the body is moved from the crime scene. This can disturb the life cycle and normal growth of the insects on the corpse. Forensic experts can detect this change in growth and find out if the body has been moved.

🪰 **Most insects** that live on corpses may be insects that are locally found at the scene of the crime. However, if the body contains insects that do not normally live in that location, the detective can be sure that the crime was committed elsewhere.

Experts can also find out other important clues about the corpse from the insects that live on the body. For example, if someone takes illegal drugs, their body undergoes many changes. These changes can affect the insects living on the body.

Flies, beetles, and gnats are some of the most common insects found on corpses. Small spiders and worms can also be spotted on dead bodies.

Forensic entomology is not a new branch of science. Back in the 1300s, the Chinese studied insects to find clues about crimes.

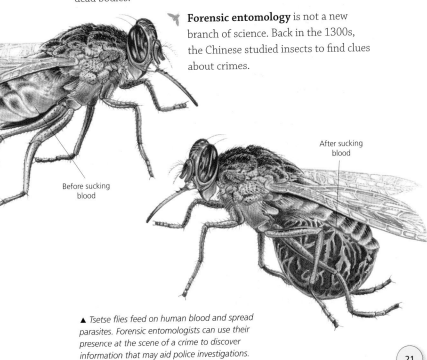

Before sucking blood

After sucking blood

▲ Tsetse flies feed on human blood and spread parasites. Forensic entomologists can use their presence at the scene of a crime to discover information that may aid police investigations.

Anatomy

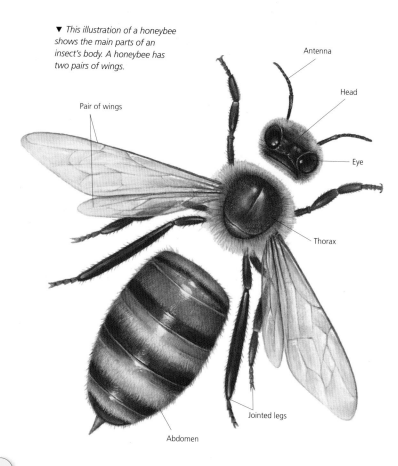

▼ This illustration of a honeybee shows the main parts of an insect's body. A honeybee has two pairs of wings.

Antenna

Head

Pair of wings

Eye

Thorax

Abdomen

Jointed legs

- **The segmented body of an insect** is divided into three parts: head, thorax (middle section), and abdomen (rear section).

- **All insects have six legs** that are joined to the thorax. They usually have either one or two pairs of wings, which are also joined to the thorax.

- **Insects have** an exoskeleton, which is a strong outer skeleton. The muscles and delicate organs of insects are enclosed and protected within this exoskeleton.

- **Two antennae** on the insect's head are used to sense smell, touch, and sound.

- **The head** also contains mouthparts that are adapted to different feeding methods, such as chewing, biting, stabbing, and sucking.

- **Insects' digestive and reproductive** systems are contained in the abdomen.

- **Insects have an open circulatory** system without lots of tubes for carrying blood. The heart of an insect is a simple tube that pumps greenish-yellow blood all over the body.

- **Special openings** on the side of the body called spiracles are used by insects to breathe.

- **Insects have a tiny brain**, which is just a collection of nerve cells fused together. The brain sends signals to control all the other organs in the body.

Molting

- **Molting** is the shedding of the hard exoskeleton periodically because it does not stretch as the insect grows bigger. All insects molt during the early stages of their life.

- **To molt,** insects swallow a lot of air or water or use blood pressure to expand their body. The exoskeleton splits and the insect emerges.

DID YOU KNOW?
A caterpillar grows about 2,000 times bigger than its size at the time of its birth. If a 7-pound human baby grew at the same rate, the baby would weigh as much as a bus in a month.

- **A soft new exoskeleton** is exposed when the insect gets rid of its old one. The new exoskeleton is bigger and allows the insect to grow.

- **The new** exoskeleton hardens and becomes darker in color.

- **Insects normally** molt five to ten times in a lifetime, depending on the species.

- **A silverfish** can molt up to 60 times in a lifetime.

- **The larval stage** between molts is known as an instar.

- **Molting takes a long time** and the insect is vulnerable to predatory attacks during this period. Most insects molt in secluded areas.

▶ *A young adult dragonfly emerges during its final molt. After resting, it will pump blood into its short, crumpled wings to spread them out to their full size.*

Communication

Insects use various methods to communicate with members of the same species, as well as with other animals. They "tell" each other about new food sources and even communicate their likes (in the case of mating partners) or dislikes (in the case of enemies).

Insects normally use their sense of touch to communicate with each other.

▲ Ants use their antennae and sense of touch as a means of communication.

Ants release pheromones (special chemical scents) in order to communicate with other ants.

Some insects use distinct colors to let other animals know that they are dangerous. Other insects may have designs on their bodies that warn predators away.

Some cockroaches and butterflies have huge, eyelike motifs on their thorax or wings, which scare predators.

Certain beetles emit strong-smelling substances from their bodies. This is to warn predators to avoid eating them. Most of the time, these strong-smelling bugs taste bad as well.

Strange colors and special scents are not always used to warn enemies. Insects also use colorful displays to attract mates.

Insects were the first animals to use sound as a means of communication. Bees use buzzing sounds to warn about danger, indicate the presence of food, and convey various other information.

The rate at which fireflies blink light is very important for successful mating. If a female firefly blinks back too fast, she is thought to be unattractive, and if her response is too slow, she is assumed to be uninterested in the male firefly.

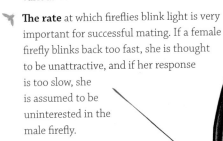

▶ Red and black colors, like those on this postman butterfly, are common among poisonous insects. Predators soon learn to avoid eating these poisonous butterflies.

Pollinators

Insects play a major role in the pollination of flowers. In fact, there are many plants that depend on insects for pollination.

Insects visit flowers for pollen and nectar. They are less likely to visit flowers that do not produce sweet nectar. For example, bees rarely visit roses because these flowers do not produce nectar.

Plants adapt their appearance to make them look attractive to insects. They reward their insect helpers with sugary nectar.

Some flowers smell sweet. This is to attract insects to come close to them. Butterflies and other nectar-sucking insects help in pollinating these flowers by carrying pollen grains on their hairy bodies.

Not all flowers smell nice. Some flowers, such as arum flowers, smell like dead and decaying matter. This is to attract flies and other pollinating insects that lay eggs in rotting matter.

Brightly colored flowers attract a lot of insects toward them. Experts have found that butterflies are most attracted to red and purple flowers, while moths are usually attracted to flowers that are pink or white in color.

Some flowers have strong lines and marks on their petals. These marks can only be seen by insects and are not visible to the human eye. They are like the landing lights at an airport and lead to the center of the flower, where the pollen and nectar are situated.

There are some species of plants in which the color of the flower changes after an insect has pollinated it. The new color discourages other insects from visiting the flower.

Insects are sometimes fooled by the appearance of plants. Some plants resemble female insects. Male insects are attracted to these flowers, and when they try to mate, the flower is pollinated.

Bees and butterflies are the most common pollinators of flowers. Beetles also help in pollination, but they are not as effective.

▼ Plants developed flowers in order to attract the insects they needed to carry their pollen from flower to flower. The partnership between insects and flowering plants has helped both groups to survive in increased numbers.

Camouflage and mimicry

Hornet moth

Hornet

◄ *The hornet moth is a mimic of a type of wasp known as a hornet that has a painful sting.*

🦟 **Insects use certain defense strategies**, such as camouflage and mimicry, to protect themselves from predators. Killer insects sometimes use the same strategies to catch their prey.

🦟 **Certain insects cleverly hide** themselves by blending in with their surroundings. This is known as camouflage.

🦟 **Some harmless insects mimic** (imitate) harmful insects in appearance and behavior. This fools predators into leaving them alone.

🦟 **Hoverflies** have yellow and black stripes on their bodies, which makes them resemble stinging insects called wasps or hornets. Predators avoid the harmless hoverflies, assuming them to be wasps or hornets.

🦟 **The larvae of some butterflies** resemble bird droppings, or even soil.

🦟 **Stick insects and praying mantises** appear to be the twigs and leaves of plants. Predators often miss out on a possible meal because these insects blend into their environment very well. Even the pupae of some butterflies look like twigs.

Monarch butterflies are bitter-tasting and poisonous, so birds do not eat them. Viceroy butterflies have orange-and-black wings similar to those of monarch butterflies. Birds avoid viceroy butterflies because they think that they are poisonous as well.

Some moths imitate dangerous wasps and bees in behavior and sound. Their buzz startles predators, which leave them alone.

Adults and caterpillars of some moths and butterflies have large, eyelike spots to scare away predatory birds.

The hornet moth has transparent wings and a yellow-and-black-striped body, making it look like a large wasp called a hornet. It even behaves like a hornet when it flies. Predators, such as birds, avoid hornet moths because they look as if they might sting.

▼ *This green leaf insect is almost impossible to tell apart from the real green leaves it is sitting on. Its large, flat abdomen looks like a leaf and even its legs look like broken pieces of leaf.*

Defense

Insects use different strategies apart from camouflage and mimicry to protect themselves from predators.

▼ *This katydid could escape from the formidable claws of a desert scorpion by shedding a leg and flying away.*

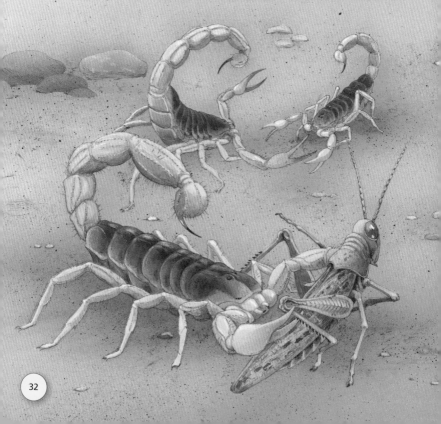

- **Some caterpillars** and larvae have special glands that secrete poison when they are attacked. Predatory birds soon learn to avoid them.

- **Stick insects and weevils** are known to "play dead" when attacked. They simply keep very still and the attacker leaves the insects alone because most predators do not eat dead prey.

- **Ants, bees, and wasps** can deliver painful stings to an attacker. These insects pump in venom and cause pain and irritation.

- **Some butterflies**, such as the monarch butterfly, are poisonous and cause the attacking bird to vomit if it eats the butterfly.

- **Certain insects** are able to shed their limbs if an attacker grabs them. This phenomenon is known as autotomy.

- **The bombardier beetle** has special glands at the end of its abdomen, which can spray hot, poisonous fluids at an attacker.

- **Some moths**, grasshoppers, and mantises suddenly show the bright colors on their hind wings to startle a predator. These are called flash colors.

- **Stink glands** present in some bugs release repelling smells that predators cannot tolerate.

DID YOU KNOW?

When alarmed, ants raise their abdomen. This sends a signal to other ants in the colony, and all the other ants raise their abdomens too.

Parasites

Organisms that live on or in other animals (their hosts) are known as parasites. Insect parasites that live on other insects are often called parasitoids.

Most parasitic insects belong to the orders Diptera and Hymenoptera. These orders include flies, ants, bees, and wasps.

DID YOU KNOW?
The eggs of the eucharitid wasp hatch into tiny larvae. These larvae attach themselves to leaves that are carried by worker ants to their nests. The wasp larvae then feed on ants in the nest.

Parasitic insects usually lay eggs in the host animal, which may be many times bigger than the parasite. The eggs hatch and the larvae (maggots are larvae) feed on the host's body.

▼ Many parasitic worms live in the gut, which is warm, safe from predators, and full of partially digested food.

| 1 Flukes |
| 2 Whipworms |
| 3 Pinworms |
| 4 Tapeworm |
| 5 Hookworm |

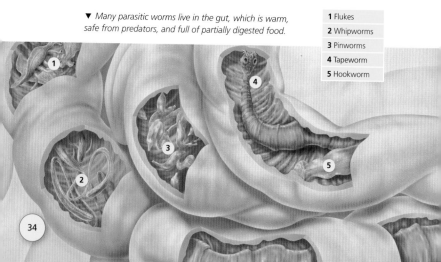

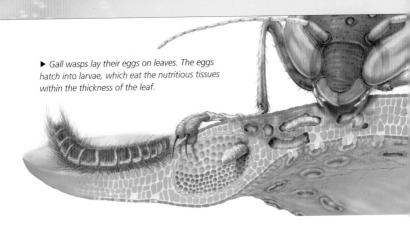

▶ Gall wasps lay their eggs on leaves. The eggs hatch into larvae, which eat the nutritious tissues within the thickness of the leaf.

Parasitoids normally live only on specific animals and insects. For instance, blister beetle larvae prefer honeybee eggs.

Parasitic larvae normally eat their way out of the host's body. They may not kill the host immediately.

The most successful parasites are those that do not kill their hosts. This ensures that they have a ready food supply for themselves and the next generation.

Stylop larvae have bristles and are long-legged. Bees pick up the larvae from flowers and take them to their nests. The larvae penetrate the bee larvae and live as parasites, first within the larvae and later in the adult bees.

Parasitic insects are often used to control pest insects. These insect parasites are beneficial to humans.

Some insects, such as lice, live on the human body and feed on blood. They cannot kill us, but lice can cause itching and irritation on the skin. Many of these parasites are carriers of diseases.

Stings

Insects that belong to the Hymenopteran order, such as ants, bees, and wasps, are the best-known stinging insects.

A stinging insect has special organs that secrete venom (poison) and a sharp sting or teeth to inject venom into the victim.

The venom can have a paralyzing effect on the prey. It can also damage tissues and cause pain. Hornet venom is the most potent.

Insect venom consists of enzymes, proteins, and chemicals known as alkaloids.

Some insects, such as mosquitoes, do not sting. They puncture the skin's surface in order to suck up blood. These insects can spread diseases such as malaria.

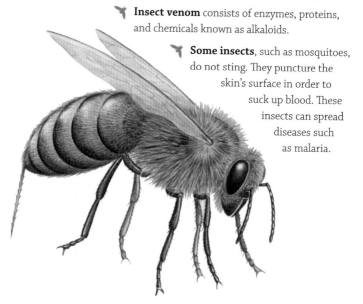

▲ A honeybee's jagged sting is a modified egg-laying tool, so only female bees can sting. Male bees do not have the necessary equipment at the end of their abdomen.

Insects sting for two purposes: to catch prey and to defend themselves from predators.

Honeybees sting only once and die soon after. The jagged stinger remains stuck in their victim's skin, which tears out the honeybee's insides.

Wasps sting their victims many times over because their stingers are smooth and can be pulled out of the victims and used again.

▼ Ants normally sting as well as bite. They inject formic acid when they sting.

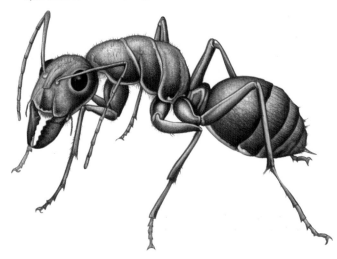

Flight

Insects fly to escape from danger, to find suitable mates, to hunt for food, or to find a new place to lay their eggs. Some insects fly very long distances in search of food, while others cross continents when they migrate.

The wing helps to distinguish between insect groups. In fact, the orders to which most insects belong are named after their wings. The order names usually end with *-ptera*, which means "wings."

Most insects have evolved two pairs of wings, which enable them to fly. The earliest insects had wings that helped them to glide through the air. These wings evolved into sturdier wings that could be flapped up and down.

Not all insects can fly. Some insects, such as female vaporer moths, have lost their wings over the course of evolution.

The size of the wing does not determine how good the insect is at flying. Some insects, such as dobsonflies, have large wings but are poor fliers.

Insect wings are membranelike structures. They have veins and nerves running across them, through which blood and oxygen are circulated. The wing's edge is usually thicker and sturdier than the rest of the wing. This helps the wing to cut through the air during flight.

The wings are attached to the insect's thorax by strong muscles. These muscles help the insect to flap its wings.

🦟 **In insects** such as beetles, the front pair of wings is hard and protects the hind wings, which are more delicate. In flies, the hind wings are reduced to knoblike halteres, which help the insect to balance itself in the air.

🦟 **A dragonfly** flaps its wings in a figure-eight pattern in the air. This movement helps to create air currents near the wings that balance the dragonfly while flying.

🦟 **Some of the fastest** and highest insect fliers belong to the order Lepidoptera—butterflies and moths. This order also has fliers that are the slowest wing-flappers.

▲ Like many other flying insects, the larvae stage of the dragonfly is spent without the benefit of wings. As adults, however, they become skilled and speedy fliers, enabling them to catch prey, such as this fly, with ease.

Habitats

Insects have adapted to survive in almost every habitat on earth, including some with extreme climates.

Entomologists have discovered certain species of insects that live on volcanic lava and others that survive in cold polar regions.

Most insects live in tropical regions, where the warm temperatures are most suitable for their growth and development.

Insects can live in freshwater ponds, lakes, streams, rivers, and even muddy pools and small water holes.

Some insect species can live very deep underwater, while others need to come to the surface to breathe in air.

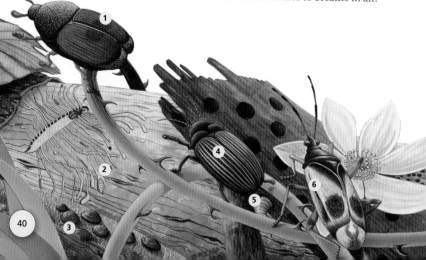

Many insects lay their eggs in water and their larvae thrive underwater. These insects fly out to live in the air when they become adults.

Some insects, such as the larvae of houseflies, live on different food materials.

Certain insects can survive on the surface of ponds of crude oil. They feed on other insects that fall into the oil.

Other insects can survive on various man-made materials, such as glue, paint, clothes, and paper. They make their homes in our homes.

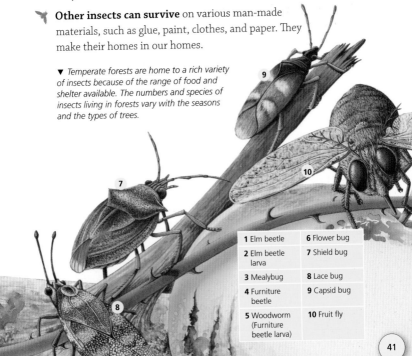

▼ *Temperate forests are home to a rich variety of insects because of the range of food and shelter available. The numbers and species of insects living in forests vary with the seasons and the types of trees.*

1 Elm beetle	6 Flower bug
2 Elm beetle larva	7 Shield bug
3 Mealybug	8 Lace bug
4 Furniture beetle	9 Capsid bug
5 Woodworm (Furniture beetle larva)	10 Fruit fly

Migration

Insects migrate across great distances in search of suitable living and breeding conditions. They often migrate when the weather gets colder or hotter, or when food becomes scarce.

Migrating insects may land in certain places and lay a large number of eggs before moving on.

Insects migrate in two ways, which are known as homeostatic and dynamic migration.

◀ Migrating locusts grow broader shoulders and longer wings to help them fly fast and for long distances. A swarm of locusts may contain billions of individuals and fly as much as 2,000 miles in a year.

▶ *Every autumn, huge numbers of monarch butterflies fly from cold northern areas of North America to the warmth of Florida, California, and Mexico. In spring, new generations of monarchs make the long return journey north.*

Homeostatic migration is when insects pass through a defined path and also return the same way. In dynamic migration, insects depend on the wind or tides to decide their path of movement.

Monarch butterflies are known to migrate across continents. These butterflies can cover a distance of more than 2,000 miles.

Butterflies may travel in huge groups of millions of butterflies. Most of the older butterflies cannot withstand the journey and die on the way.

Pilots have spotted migrating butterflies at an altitude of 4,000 feet.

Locusts migrate across farmlands in large swarms in search of greener pastures.

Army ants do not build permanent nests because they constantly migrate in search of food.

Ants

- **Ants are one of the most successful insects** on earth. There are more than 9,000 different species.

- **The study of ants** is known as myrmecology. People sometimes rear ants in elaborate ant farms.

- **Ants belong** to an order of insects known as Hymenoptera. Bees and wasps also belong to this group. Entomologists believe that ants evolved from wasps millions of years ago.

- **Ants are social insects**, which live in huge colonies. These colonies consist of the queen ant, female workers, and male ants.

- **In ant colonies**, the different ants divide themselves into groups that perform various tasks. Some ants are cleaners, some take care of the young ones, and others gather food or defend the nest.

- **When ants find food**, they form a chemical trail of pheromones so that other ants can find their way from the nest to the food.

- **Rare trap-jaw ants** are found in Costa Rica. They have large jaws, which are used to trap and stab the enemy.

- **An ant can lift a weight** that is 20 to 50 times more than its own body weight.

- **Slave-making ants** do not have their own nests. They steal the pupae of other ants and make them slaves in their own colony.

DID YOU KNOW?

Birds allow ants to crawl on their bodies and spray them with formic acid. Known as anting, this gets rid of parasites from their feathers.

▲ Aphids produce a sugary
substance that ants feed on.
In return, the ants protect the
aphids and "graze" them in fresh
pastures.

47

Ant facts

Most ants live in anthills that are made of mounds of soil, sand, and sticks, but some nest in trees.

Anthills consist of different chambers and tunnels. The chambers are used for different purposes, such as food storage, nurseries, or resting areas.

▼ *Ants communicate with the help of their feelers (antennae). They also leave scent trails behind them to let the other ants know exactly where to find the food source.*

Not all ants live in anthills. Army ants are nomadic insects. They carry their eggs and young ones along with them while traveling, and set up temporary camps.

The largest ant in the colony is the queen ant. When she matures, the queen ant flies off in search of a suitable place to build a new colony.

Queen ants nip off their wings once they find a place to breed. Smaller worker ants also have wings.

Worker ants take good care of the eggs. At night, they carry the eggs into deep nest tunnels to protect them from the cold. In the morning, workers carry the eggs back to the surface to warm them.

Male ants die shortly after mating with the queen ant.

Ants have two stomachs. One stomach carries its own food while the other carries food that will be shared with other ants. This is called the crop.

Each ant colony has a unique smell that helps the members to identify each other. This also helps the ants to detect an intruder in the nest.

Ant nests

Ants live in nests that are built by the worker ants. These nests are known as formicariums.

Ant nests are usually built on the ground. These nests can be either underground or above the soil. Some species of ants support their nests by building them against man-made structures, such as telephone poles.

▼ Ants can build large and complex nests. This is due to their social lifestyle and great strength—a single ant may lift an object 20 times its own body weight.

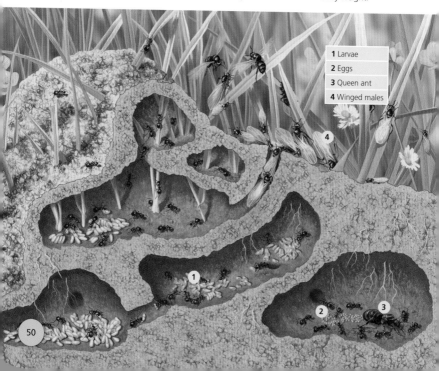

1 Larvae
2 Eggs
3 Queen ant
4 Winged males

DID YOU KNOW?
The pagoda tree ant builds its nest with wood pulp and saliva. These ants place small scraps of tree bark one on top of the other, making a tall structure that looks like an Asian pagoda.

- **Ant nests that are built** above the ground are in the form of mounds of soil, twigs, and leaves. Some ant mounds can be as tall as 3 feet. These ant nests are several feet wide.

- **Different species** of ants build different kinds of nests. Some nests are simple, with only a few chambers and corridors, while other nests are complex and have separate chambers for the queen, the eggs, and the young.

- **Worker ants** take care to control the temperature inside the nest and ensure that there is proper ventilation in the chambers. The walls of these chambers are made smooth with mud and saliva.

- **Some species** of ant grow "fungus gardens" inside their nests. These gardens provide food for the young ants. There are special cells inside the nest used to store food material for the fungus.

- **Certain ants** build nests in trees. These nests are not confined to a single tree. The ants connect the different trees with the help of underground tunnels and corridors.

- **Leaf-cutter ants** make the largest ant nests. Some leaf-cutter nests contain millions of worker ants.

- **Ants sometimes make use** of old, empty wasp nests. They also live in tree galls, dead wood, and tree stumps.

Army ant

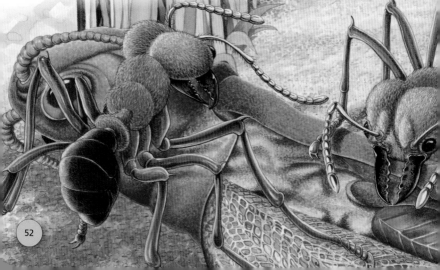

Army ants constantly migrate in search of food. They can attack and enslave ants living in other colonies. Army ants march at night and stop to camp in the morning.

Nomadic in nature, army ants do not build permanent nests.

Temporary nests are formed by army ants while the queen lays her eggs. The ants cling onto each other and form the walls and chambers.

Army ants are voracious eaters. They march in swarms of up to one million ants and eat almost 50,000 insects a day!

▼ The army ants of tropical America march in columns, just like real soldiers. To cross gaps, some of the ants form bridges with their bodies, allowing the rest of the army to swarm over the living bridge. The worker ants also link up to form chains that surround the queen and young.

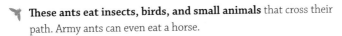

These ants eat insects, birds, and small animals that cross their path. Army ants can even eat a horse.

Army ants have simple eyes, not compound eyes like other ant species. However, worker army ants are blind.

Army ants have not evolved (changed) much in the last 100 million years.

Some people allow army ants to march into their homes and clear them of insects and other pests.

DID YOU KNOW?
A queen army ant can lay up to four million eggs in one month.

53

Leaf-cutter ant

🐜 **Leaf-cutter ants** cut out bits of leaves from plants and carry them back to their underground nest.

🐜 **These ants cannot digest leaves**. Leaf-cutter ants feed on a fungus that is specially grown by them.

🐜 **The cut leaf pieces** are used to fertilize special fungus farms that are grown inside the ant nest. There can be numerous fungus farms in a single nest.

🐜 **Leaf-cutter ants** are normally found in tropical rain forests.

🐜 **A queen leaf-cutter ant** can produce about 15 million offspring during her lifetime.

🐜 **Leaf-cutter ants divide** themselves into workers and soldiers. The biggest ants are soldiers, which protect the nest.

▶ Leaf-cutter ants carry cut pieces of leaf back to their nest. Each leaf fragment can take two or three minutes to cut and is often many times the size of the ant.

- **Small worker ants** take care of the young and manage the nest, while bigger worker ants go out and cut pieces of leaves for the fungus farms.

- **In some species** of leaf-cutter ants, tiny workers ride on the pieces of snipped leaves. They protect the larger workers from flies that try to lay their eggs on them.

- **In some cultures**, leaf-cutter ants are eaten. They are a rich source of protein.

- **Leaf-cutter ants do not sting**, but they can bite.

Weaver ant

🐜 **Weaver ants** build their colonies in the tops of trees, using live green leaves.

🐜 **The larvae of weaver ants** secrete a sticky silklike substance. Adult ants use the young larvae like glue sticks.

🐜 **A team of adult worker weaver ants** holds two leaves together, while a single worker holding a larva runs through the edges and "sews" them together. The worker ant holds the larva in its mandibles (mouthparts) and uses the silky secretion to stick the leaves together.

🐜 **A colony can contain** about 150 weaver nests in 20 different trees. The queen ant's nest is built in the center of the colony. It is made with extra silk and is feathery in appearance.

🐜 **The larger worker (soldier) ants** fiercely protect their nest, while the smaller workers take care of various chores inside the nest.

🐜 **Weaver ants are carnivores** and feed on body fluids from small, soft-bodied insects. Some species also feed on honeydew.

🐜 **For 2,000 years**, the Chinese have used weaver ants to control pest infestations in their crop fields.

DID YOU KNOW?
Weaver ants are eaten in Eastern cultures. Adult ants have a lemony taste and are used to flavor rice. The pupae are said to be creamy in taste. The oil extracted from these ants is used as a sweetener.

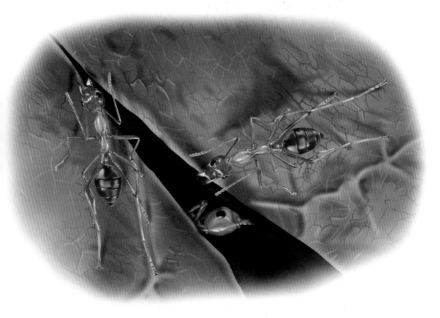

▲ *Weaver ants join leaves together to make their nests.*

Weaver ants do not sting, but they can inflict very painful bites if provoked. When they bite, weaver ants squirt formic acid into the wound, causing even more pain.

Some caterpillars and spiders camouflage themselves as weaver ants and attack weaver ant nests. The spiders may even smell like weaver ants.

Bees

- **There are approximately 20,000** species of bees. Many bees live alone, but over 500 species are social and live in colonies.

- **Bees are small** in size, ranging from 1–2 inches in length. They have a sting to protect them.

- **Generally black or gray** in color, bees can also be yellow, red, green, or blue.

- **Bees feed on pollen and nectar** collected from flowers. Hairs on their body help them to collect the pollen. Pollen contains protein and nectar provides energy.

- **Social bees secrete wax** to build their nests. A honeybee colony may contain 3,000 to 40,000 bees depending on the species, the season, and the locality. It consists of a single queen bee, female workers, and male drones.

- **Male drones** do not have stingers and their function is to mate with the queen bee. The queen lays about 600 to 700 eggs every day.

- **Bees have five eyes**. They have two compound eyes and three simple eyes, or ocelli. Bees can see ultraviolet light, which is invisible to the human eye.

- **A normal bee's life span** ranges from five to six weeks, but a queen bee can live for up to five years.

- **Bees communicate** with other bees about the distance, direction, and quality or quantity of the food source through a unique combination of dance and sounds.

A honeybee attacks either to protect itself or its colony. Once a bee stings, it leaves behind its stinger and venom in the victim's body. As the bee pulls away from the victim, its organs are pulled out of its abdomen, killing the bee.

▼ *Honeybee workers crowd around their queen. The workers lick and stroke their queen to pick up powerful scents called pheromones, which pass on information about the queen and tell the workers how to behave.*

Bee behavior

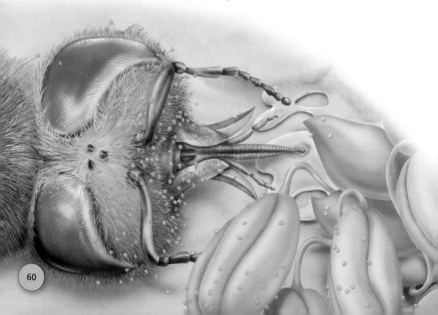

Bees have two kinds of mouthparts. The first kind, found in honeybees, is adapted for sucking. The other kind is adapted for biting. This is found in carpenter bees.

The antennae are the organs of touch and smell. Bees use their antennae to detect flower fragrances and to find nectar.

Bees can rarely distinguish sweet and bitter tastes but can identify sour and salty tastes. Bees use their front legs, antennae, and proboscis for tasting.

▼ This bee's long, pink, ridged glossa (tongue) laps up nectar from deep in the flower. Bees are well equipped to eat both solids and liquids, and to fashion wax to make cells for the honeycomb.

- **Bees have no sense of hearing**, but they can sense vibrations through their sensitive hairs.

- **Bees do not sleep**, although they may remain motionless.

- **Wild bees nest underground** or in tree holes, caves, or under houses. Honeybees also live in hives constructed by people.

- **Social bees** follow a hierarchical structure. They live in large colonies of queens, males, and workers. The queen's cell is structurally different from the workers' cells. Males do not help in the organization and other activities of the colony.

- **When a colony** becomes overcrowded, some bees fly to a different location. This phenomenon is called swarming. It is a part of the annual life cycle of the bee colony.

- **A division of labor** exists among social bees. The queen bee lays eggs and male bees fertilize the queen bee. Worker bees perform various tasks, such as cleaning the cells, keeping the young warm and guarding them, feeding larvae, producing wax, and collecting food for the colony.

- **The queen bee** secretes pheromones, which tell worker bees that she is alive and well and also inhibit the development of worker bees into queens. Once she lays eggs, the fertilized eggs become female worker bees and the unfertilized eggs become male bees.

Bee facts

 Bees belong to the order Hymenoptera. Ants and wasps also belong to this order.

 Bees look like wasps but they have more hair and thicker, more robust bodies. Unlike wasps, bees have specialized organs to carry pollen.

DID YOU KNOW?

Bees cannot see the color red. They are mainly attracted to blue and yellow flowers and visit these flowers more often.

 The most important group of pollinating insects is formed by bees. People have used bees since prehistoric times to pollinate crops.

 Entomologists distinguish between bee species and families by studying the subtle differences in the veins of their wings, their body structures, and their nesting and feeding habits.

 The richest diversity of bees is found mostly in the northern regions of Mexico and southern Arizona.

 Solitary bees do not live in large colonies and instead build their own nests where they rear their offspring.

 Some of the best-known solitary bee species are carpenter bees, leaf-cutter bees, and mason bees.

 The popular honeybee species *Apis mellifera* is commonly found in Europe, Africa, Asia, the Middle East, and the Americas. The scientific name of the honeybee means "honey carrier."

Bees are very hardworking insects. The term "busy as a bee" comes from their tireless way of collecting pollen and nectar from flowers.

Fossil records show that the first bees appeared in the Miocene epoch some 26 million years ago. These were the leaf-cutter bees.

▲ Bees are important because they carry pollen from flower to flower so that seeds can grow. The bright colors of flowers and the nectar they produce encourage bees to visit them.

63

Bee nests

Social bees make complex nests, which consist of a number of cells, often built in flat sheets called combs.

Social bees, such as honeybees, bumblebees, and stingless bees, make cells of wax. However, honeybees are the only species that makes prominent honeycombs. People sometimes keep honeybees in beehives for their honey.

Honeycombs are made up of hexagonal cells divided into three main sections. The upper section is used for storing honey and the middle section for storing pollen. The lower section is used to house the eggs and young.

The hexagonal shape of the honeycomb cells allows the maximum amount of honey to be stored and uses the least amount of wax.

Social bees use their nests to raise their eggs, larvae, and pupae, collectively called their brood. Some also use their nests to prepare and store honey for the winter.

Bees use "bee glue" (propolis), a sticky tree resin, to strengthen and repair their nests.

To maintain an optimum temperature of 95°F in the beehive, bees use water as a coolant. They also flutter their wings to maintain the temperature at the correct level.

Stingless bees build saclike combs. They are made from a mixture of resin and wax called cerumen. The combs are held together by propolis in the hollows of trees, rocks, and walls. While building their nest, stingless bees hang the cells horizontally instead of aligning them vertically, as honeybees do.

A bumblebee queen builds her nest in a hole in the ground. She may use the abandoned nests of birds, mice, ants, or termites.

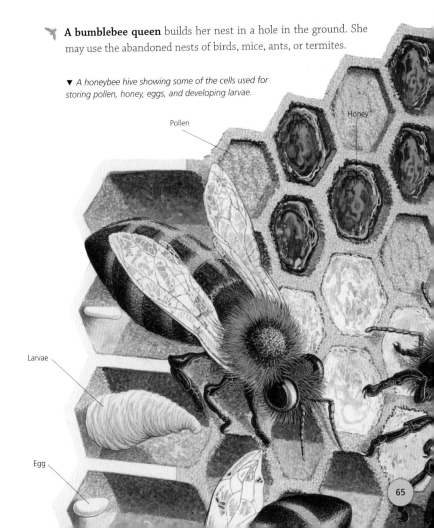

▼ *A honeybee hive showing some of the cells used for storing pollen, honey, eggs, and developing larvae.*

Pollen

Honey

Larvae

Egg

Honey

▶ *A natural honeycomb, showing the double layer of six-sided cells constructed by the bees to store honey. The honeycomb is used in the winter as food for the larvae and other members of the bee colony.*

Honey is a natural, unrefined sweetener, which is an alternative to sugar.

One ounce of honey contains approximately 90 calories.

Bees collect the nectar of flowers, along with other plant secretions, and turn it into honey. It is then altered chemically into different types of sugars and stored in the comb cells.

The honey in a honeycomb includes matured nectar, pollen, bee saliva, and wax granules.

Honey mainly contains water, sugars, and minerals. Trace elements, such as calcium, phosphorous, magnesium, iron, silica, and vitamin C, are also present.

Honey is considered to be the only source of food that has all the energy and protein reserves necessary to sustain life.

The color and flavor of honey depends on the climate and the flowers from which the nectar has been collected.

● **Honey serves** as an important source of medicines because of its mild laxative, bactericidal, sedative, antiseptic, and alkaline characteristics.

● **Honey extracted** from wild beehives can be dangerous if the nectar is obtained from poisonous flowers.

● **Honey** is used in a fluid medium in the preservation of the cornea (layer at the front of the eye).

▶ *When they remove the honeycombs from a hive, beekeepers have to wear special clothes and netting over their face to protect their skin from bee stings.*

67

Honeybee

Honeybees are social insects. They live in large colonies and are the most popular species of bees.

Some well-known species of honeybees are Italian bees, Carniolan (Slovenian) bees, Caucasian bees, German black bees, and Africanized honeybees.

In a honeybee colony, different groups of bees carry out different tasks. A colony is made up of a queen bee, female workers, and male drones.

▼ Stored honey keeps worker honeybees and their developing young alive during the long, cold winters in temperate areas when other bee colonies die out.

▲ *Honeybees have hairs on their body to help them collect pollen.*

- **Queen bees** are sexually productive and responsible for laying eggs that develop into drones and worker bees.

- **Workers have a sharp stinger**, pollen baskets, wax-secreting glands, and a honey sac for collecting honey. The wax is used to build sheets of cells called combs.

- **Worker bees build** the nest as well as collect pollen and nectar for food. They are also responsible for maintaining the nursery temperature at 95°F, which is ideal for hatching the eggs and rearing the larvae.

- **Drones do not have a stinger**. Their sole function is to mate with the queen. They die after mating. During the winter, they are driven out of the beehive by the workers.

- **The biggest producers of honey** are honeybees. This is why they are the species most domesticated by humans.

- **Honeybees** have an amazing mode of communication among themselves—dancing. Dr. Karl Von Frisch won the Nobel Prize for deciphering the bee dance.

- **Honeybees are susceptible** to various diseases and attacks by parasites. Parasite and virus attacks may cause paralysis in bees.

Killer bee

🐝 **Africanized honeybees** are popularly known as killer bees among entomologists and beekeepers. These bees are very aggressive and attack their enemies at the slightest chance. This is how they got their name.

🐝 **Brazilian scientists** first brought the ancestors of Africanized honeybees from Africa to Brazil in the 1950s. These bees were called African honeybees.

🐝 **Scientists bred** these African honeybees with European ones in order to increase the production of honey. The new species of bee was named the Africanized honeybee or killer bee. These bees adapted well to the climate and their population increased rapidly.

🐝 **Killer bees** can nest in natural surroundings, such as hollow trees, as well as man-made surroundings, such as porches, sheds, and even garbage containers.

🐝 **These bees** feed on nectar and pollen and make honey from the nectar.

🐝 **Killer bees** can survive on sparse supplies of nectar. Therefore, they produce and store less honey than honeybees.

🐝 **These bees** are known for their frequent swarming. They migrate to new colonies when the climatic conditions are not suitable.

🐝 **Beekeepers** find it very difficult to manage these bees because of their rapid population growth. However, killer bees are useful for crop pollination.

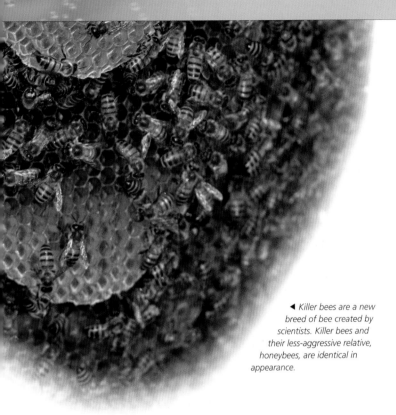

◄ Killer bees are a new breed of bee created by scientists. Killer bees and their less-aggressive relative, honeybees, are identical in appearance.

Killer bee colonies can easily become agitated and pose a serious threat to humans and other animals. When a hive of killer bees is disturbed, they can viciously chase their enemies for a considerable distance and the nest can stay disturbed for days. These bees sting their enemies collectively.

71

Bumblebee

Bumblebees are hairy, black, yellow, or orange in color and up to one inch long. They are most common in temperate regions and are less aggressive than other bees.

Bumblebees live in small colonies of between 50 to 600 bees.

▼ Bumblebees collect pollen from flowers in pollen baskets (long, stiff bristles) on their back legs. The bee combs the yellow pollen dust from its body and packs it tightly into the pollen baskets until they are full.

Bumblebees build their nests on the ground in rocky holes, grassy hollows, or deserted rodent or bird nests. The chambers are spherical in shape with one exit. The cells inside are shaped like capsules.

DID YOU KNOW?
Bumblebees do not produce large quantities of honey. They store honey just for feeding themselves and their young.

The queen bumblebee lays her eggs in wax cups inside the nest. The wax is secreted by her abdominal glands.

Only the young queen bumblebees survive the winter. The rest of the bees in the colony die.

Bumblebees regulate their body temperature with the help of their body hair. The queens hibernate in the winter and come out in the spring to lay their eggs and start a new colony.

Low fliers, bumblebees move slowly around the flowers.

Bumblebees help to pollinate plants such as red clover. Their long tongue enables them to reach deep inside flowers.

In summer, the bumblebee workers fan their wings to cool the developing young. The buzzing sound of fanning is so loud that it can be heard from some distance away.

◄ *There are different species of bumblebees, such as this buff-tailed bumblebee, which is found in Europe.*

73

Cuckoo bee

Bees that are parasitic are popularly known as cuckoo bees.

There are two types of parasitic bees—cleptoparasitic and social parasitic. Cleptoparasitic bees invade the nests of solitary bees, hide their eggs in the brood chambers before the hosts can lay theirs, and then close the chambers. Social parasitic bees target the resident queen and kill her. Then they lay their own eggs in the host's cells, and force the host's workers to raise the young parasitic bees.

Cuckoo bees are found in a variety of habitats, such as gardens and grasslands.

These bees do not forage for food and cannot provide for their young ones because they do not have pollen baskets on their legs. Since they do not collect pollen or rear their young ones, both these roles are carried out by the host bees.

Some cuckoo bees lay their eggs in the nests of bumblebees. Unknowingly, the bumblebees raise the emerging larvae of the cuckoo bees as their own by feeding and taking care of them.

Cuckoo bee larvae have huge jaws and use them to kill the larvae of their hosts.

There are no workers among cuckoo bees. There are only males and females.

Male cuckoo bees help in the mating process and die before the winter starts.

After mating, the females go into hibernation and sleep throughout the winter, emerging only in the spring to lay their eggs in the host's nest.

Females have a thick cuticle, which is the outermost layer of the skin. This helps to protect them from attacks by other bees when they try to invade their nests.

▼ *Laying eggs in the nests of other bees may be a successful method of reproduction for cuckoo bees, but it does carry risks because they may be killed by the potential host bees.*

75

Carpenter bee

Carpenter bees are named after their habit of drilling into wood to build nests.

Blue-black or metallic in color, carpenter bees resemble large bumblebees.

▼ A carpenter bee is about an inch long. It is not as hairy as a bumblebee, with short hairs on its abdomen or sometimes no hair at all.

- **Carpenter bees are found** all over the world, especially in areas where woody plants flourish. They are common in forested regions of the tropics.

- **Solitary in nature**, carpenter bees do not live in colonies.

- **Male carpenter bees** have white-colored faces or white markings, and females have black-colored faces.

- **Males do not** have a stinger but they do guard the nest. The females have stingers but are very docile and do not sting unless in danger.

- **Female carpenter bees** nest in their wooden tunnels. They prefer weathered, unpainted, bare wood. In these tunnels, carpenter bees drill holes, where they lay eggs in individual cells and store enough food for the larvae to grow. There is only a single entrance to each tunnel.

- **People take preventive measures**, such as spraying pesticides, to keep carpenter bees away from their homes and gardens.

DID YOU KNOW?

Adult carpenter bees hibernate during the winter. They remain in their wooden nest, surviving on stored honey and pollen.

Mason bee

🐝 **Mason bees are solitary bees**. A single bee builds its nest and rears its eggs.

🐝 **These bees** are common in the forested regions of the western United States but are found in many parts of the Northern Hemisphere.

DID YOU KNOW?
Some mason bees nest inside empty snail shells. A large shell may contain up to 20 eggs.

🐝 **Although mason bees** are generally smaller than honeybees, some species of mason bees can be slightly larger or of the same size. Male mason bees are smaller than the females, but they have longer antennae.

▼ Mason bees lay their eggs after filling the cells with enough pollen and honey for the larvae to feed on.

These **bees** are not aggressive by nature. They do not sting unless they are disturbed or mishandled.

Mason bees make their nests from the chewed paste of plant fiber or mud. Several species of mason bees build their nests in crevices or holes in wood, which is why they are called mason bees. Once the nests are built, the bees mate in the summer.

▲ *A mason bee builds its nests in wood or in the soft mortar between bricks. While not dangerous to humans, these bees can damage walls and buildings.*

A nest has approximately 6–12 cells. Mason bees hibernate during the winter and emerge from their nests in the spring.

People, particularly farmers, cultivate these bees because they pollinate crops and fruit plants.

In orchards, drilled wood is placed to attract mason bees. They are important pollinators of apple and cherry trees, as well as blueberry and other fruit plants.

Leaf-cutter bee

Leaf-cutter bees are named after their habit of cutting pieces of leaf to make a protective casing for their eggs. They nest in the soil, hollow plant stems, or woody tunnels.

Black in color, leaf-cutter bees have hair on their abdomen. This hair helps them collect pollen, unlike other bees, which collect pollen on their legs.

Most leaf-cutter bees are solitary in nature. Females like to construct individual nests independently. Males are smaller than females and have hairy faces.

The female bee builds her nest using semicircular leaf pieces for the side walls and circular pieces for the ends. Once the cell is ready, she stores pollen and honey inside, lays an egg, and closes each cell with a perfectly fitted disk of cut leaf. Then she begins the sequence again until the entire nest is complete.

Individual female leaf-cutter bees do all the work. They select the nesting place, construct cells, lay the eggs, and rear the larvae.

DID YOU KNOW?
The average life span of a female leaf-cutter bee is two months, and in this time, she lays 30 to 40 eggs.

Leaf-cutter bees are docile in nature with a mild sting. They use the sting to defend themselves from attack.

These bees can harm plants because of their habit of constructing nests with plant leaves. Their favorite plant is the rose plant.

🐝 **Leaf-cutter bees help some plants,** such as alfalfa, with pollination. They do this by carrying the pollen from one plant to another.

🐝 **Leaf-cutter bees** have many enemies, such as wasps, velvet ants, and some species of blister beetles.

▼ *A female leaf-cutter bee rolls up the leaf fragment she has cut with her sharp, scissorlike jaws and carries it between her legs as she flies back to the nest.*

Wasps

Wasps belong to the Hymenopteran order, as do bees and ants. They have a hard exoskeleton and their body is divided into head, thorax, and abdomen. They have four transparent wings and two compound eyes.

Wasps are solitary as well as social insects. Social wasps live in huge colonies while solitary wasps live alone. There are about 17,000 species of wasps, but only about 1,500 species are social.

Wasp nests can be simple or complex. Some nests are just burrows in the ground, while others are built with mud and twigs and can have many cells and tunnels.

Each nest has at least one queen wasp as well as workers and males.

Not all wasps build nests. Some wasps, such as cuckoo wasps, lay eggs in the nests of other bees and other wasps.

Other species of wasps lay their eggs in stems, leaves, fruits, and flowers instead of building nests.

◄ The small nest entrance is easy to defend from other insects and also helps the wasps to control the humidity and temperature inside the nest.

Adult wasps feed on nectar, and fruit and plant sap, while the larvae feed on insects.

Many species of wasps are parasitic in nature, which means that they live part of their lives as parasites inside other insects. The larvae of such wasps feed on other insects and sometimes eat plant tissues.

▲ *Wasps have large black compound eyes, which are good at detecting movement. They also have sharp, cutting jaws with jagged edges.*

Wasps are helpful for controlling pests such as caterpillars.

▼ *Thousands of wasps live in one nest, built with a papery substance made by the wasps. The queen wasp lays her eggs, from which worker wasps hatch. They then help to enlarge the nest.*

83

Gall wasp

Gall wasps are small parasitic insects that feed on plants.

These insects are named after their habit of causing the formation of a tumorlike growth in plants, known as galls.

▼ Female fig wasps lay their eggs on some of the flowers inside immature figs. The eggs develop into larvae, pupae, and adults. Adult females emerge and fly off to find other figs in which to lay their eggs. The wingless male wasps die without ever leaving the fig.

▶ Gall wasps are usually about 2–8 mm long. Their shiny abdomen is oval in shape and their wings have few veins.

Galls are an abnormal growth of plant tissues and leaves. Some of them look like greenish apples or berries on leaves.

When a female gall wasp injects her eggs into a plant, galls are formed. When the eggs hatch, the larvae release chemicals, which causes the plant to cover them with soft tissues in the form of a gall.

Galls can be either spongy or hollow inside.

Gall wasp larvae feed on the gall and pupate inside it. Adult gall wasps emerge from the gall either by boring a hole or by bursting through its surface.

Different types of galls, such as leaf, flower, seed, and stem galls, are caused by different species of gall wasps.

A gall is like a nursery for one or more species of gall wasp.

Fig wasps, a species of gall wasp, cause the formation of seed galls inside wild figs and pollinates them in the process. No other insect pollinates wild figs.

Gall wasps are very selective about the plants on which they lay their eggs. For instance, some gall wasps lay their eggs on figs, while others prefer roses.

Hornet

🐝 **Hornets belong** to the order Hymenoptera and the Vespidae family. They have dark brown and yellow stripes all over their body.

🐝 **These insects are known** for their ferocious nature and painful sting. They are huge, robust wasps and are social in nature.

🐝 **Social hornets** form huge colonies that can contain about 25,000 individuals.

🐝 **Hornets can build** their nest anywhere—at a height or even on the ground. They insulate their nests with layers of "paper."

▲ *The super wasps known as hornets can be up to 30 mm long. Apart from their larger size, hornets can be distinguished from smaller wasps by their deeper yellow colour.*

DID YOU KNOW?

Hornets are known to chase their tormentors. Hence the saying "never stir a hornet's nest."

🐝 **These insects chew plant fiber** and mix it with saliva to form a papery paste, which they use to build nests. The nest is spherical, with an entrance at the bottom, and is divided into many tiers inside. These tiers have hexagonal cells, in which the young are raised.

Hornets build the largest nests of all wasps. A hornet nest can be 4 feet long and 3 feet in circumference.

Hornet colonies die out in one year. No member of the colony survives the winter except the female hornets that have mated.

Abandoned hornet nests provide shelter for other insects during the winter.

Some insects have stripes that resemble those of hornets. These ward off predators, which mistake harmless insects for hornets.

▼ *Although hornets are very protective of their nest, they are far less likely to sting than smaller wasps. A hornet's nest can be as large as a basketball and is constructed in a hollow tree or under roof eaves, porches, and outbuildings.*

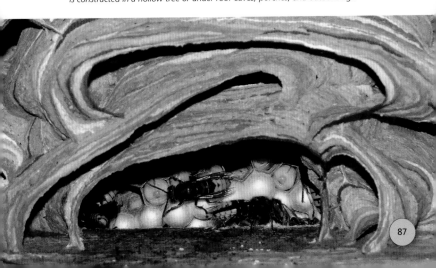

Paper wasp

- **Paper wasps are reddish-brown** in color and have yellow stripes on their body.

- **Social insects**, paper wasps live in small colonies of 20 to 30 insects. After the queen wasp mates, she builds a nest with a material similar to papier-mâché (paper pulp). It is made of six-sided cells. The nests are mostly built in the spring.

- **Paper wasps chew** plant fiber, which they mix with their saliva to build their nests.

▼ This Costa Rican paper wasp is starting to build her nest under a leaf. She will use her antennae to measure the size of the cells.

▶ Wasps do not store food in their nest, as the cells face downward and are open at the bottom. The queen wasp glues her eggs inside the cells to keep them from falling out.

Some paper wasp nests look like inverted umbrellas, which is why these insects are also known as "umbrella wasps."

A few queen paper wasps build a nest together. The most powerful queen dominates and leads the colony, while the rest become workers.

The subordinate queen wasps are called joiners. Sometimes, a joiner manages to overpower the reigning queen. She becomes the new queen while the original queen becomes a worker.

Unlike other wasps, bees, and ants, queen paper wasps closely resemble the worker wasps.

Adult paper wasps feed only on nectar, while the young larvae feed on chewed insects.

Sometimes, army ants invade paper wasp nests and destroy the entire colony.

Parasitic wasps

🐝 **Parasitic wasps lay their eggs** inside other wasps, spiders, bees, caterpillars, and aphids.

🐝 **Female parasitic wasps** inject their eggs into the body of their host with the help of an ovipositor, or egg-laying tube. Once the eggs hatch, the larvae literally eat their way out of the host.

🐝 **Parasitic wasps** can lay as many as 3,000 eggs inside a single host insect.

🐝 **Some parasitic wasps,** such as chalcid and braconid wasps, are known for infesting moth and butterfly caterpillars. They are helpful for controlling the caterpillar populations in crop fields.

🐝 **Some very tiny wasps** specialize in parasitizing insect eggs. These eggs have the ability to multiply into many cells, and almost 150 individual wasps of the same sex can hatch from a single egg.

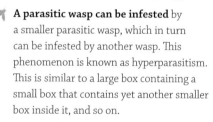

🐝 **A parasitic wasp can be infested** by a smaller parasitic wasp, which in turn can be infested by another wasp. This phenomenon is known as hyperparasitism. This is similar to a large box containing a small box that contains yet another smaller box inside it, and so on.

🐝 **Unlike** other social wasps and bees, parasitic wasps do not sting.

◀ *The smallest insect, the fairyfly, is an egg parasite wasp. A fairyfly measures only about 0.2 mm.*

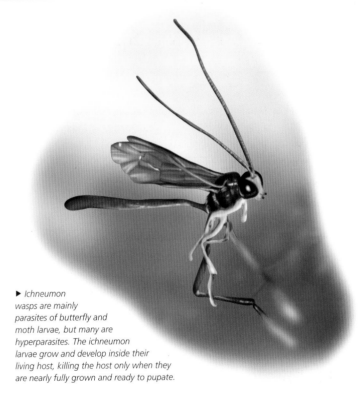

▶ Ichneumon
wasps are mainly
parasites of butterfly and
moth larvae, but many are
hyperparasites. The ichneumon
larvae grow and develop inside their
living host, killing the host only when they
are nearly fully grown and ready to pupate.

Aphids that serve as hosts for parasitic wasps appear puffy and hard. They die once the wasp larvae are ready to pupate and are then known as "aphid mummies."

Parasitic wasps are useful to us because their normal hosts are pest insects.

Velvet ant

▼ The tough outer covering of a velvet ant protects it against stings.

- **These wasps** resemble huge, hairy ants, which is why they are known as velvet ants.

- **Velvet ants** are usually red, brown, or black in color.

- **Found in dry areas**, velvet ants are densely covered with long hair.

- **When it is too hot** to venture outside, velvet ants burrow underground or climb into plants.

- **Male velvet ants** have wings and cannot sting. Females, on the other hand, do not have wings and can sting. They move about on the ground like ants and their sting can be quite painful.

- **Females** move very swiftly and are often found busily searching for the burrows of solitary bees and wasps.

- **Velvet ants** lay eggs in the nests of bees and other wasps, and are parasitic.

- **The larvae** of the velvet ant emerge before the eggs of the host hatch. They eat the host's eggs as well as its larvae.

- **Velvet ants** make a squeaky noise if they are attacked or captured.

DID YOU KNOW?
The sting of the velvet ant is so powerful that people believed it could kill a cow. This, however, is not true.

Butterflies and moths

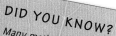

- **Butterflies and moths** are part of the order Lepidoptera, which means "scaly wings." There may be 200 to 600 tiny scales on every square millimeter of wing.

- **Moths and butterflies** are ancient insects. Fossil records show that moths date back 140 million years and butterflies 40 million years.

- **Butterflies** usually have delicate and slender bodies. Moths tend to have more plump and robust bodies.

- **One feature** that butterflies and moths have in common is that they cannot survive extreme cold weather. Not one single moth or butterfly is found in Antarctica.

- **Both butterflies and moths** feed on nectar as well as on animal fluids.

◀ Butterflies, such as this morpho butterfly, have two pairs of wings, which often have beautiful markings or are brightly colored.

▶ *Moths are more likely than butterflies to be active at night.*

- **Butterflies** and moths are important pollinators of plants.

- **Butterflies** usually fly by day, while moths are active after sunset. There are some exceptions, such as the evening brown butterfly that flies at dusk. Similarly, there are brightly colored moths, such as the blue tiger moth, that fly in broad daylight.

- **While resting**, most butterflies fold their wings straight up above their backs. Most moths keep their wings spread flat or fold them like a tent over their bodies.

- **Most moth caterpillars** weave a structure around themselves for pupation. This is called a cocoon and is made of silk, leaves, or soil. Most butterflies do not weave such cocoons.

- **The antennae** of a butterfly are plain stalks ending in a club (or a hook in the case of skipper butterflies). The antennae of moths have many patterns, ranging from single strands to feathery branches, but they never end in a club.

97

Butterfly lifestyle

Butterflies do not have a mouth with teeth to chew or break their food into small pieces. Instead they have a long, strawlike structure called a proboscis under their head, which helps them suck nectar and other juices.

Butterflies pass through four different stages of development, also called metamorphosis. First, the adult butterfly lays eggs on plants. The egg hatches into a caterpillar or larva. The caterpillar then develops into a pupa or chrysalis. Finally, the pupa matures into a butterfly.

Caterpillars grow 27,000 times bigger than their original size when they first emerge from the eggs. They shed their skin from time to time, as it does not expand with their growth. This process is called molting.

Certain male butterflies have scent pockets on their wings, which disperse pheromones, a chemical that stimulates mating.

Some butterflies migrate in order to avoid bad weather, overcrowding, or to find a new place to live.

The average life span of butterflies is 20 to 40 days, but some species can survive up to ten months while others last only three to four days.

Butterflies consume only liquid food, such as flower nectar and liquids from rotten fruits or vines. Some feed on liquid animal waste.

◄ A butterfly drinking nectar from a flower using its proboscis.

Butterflies protect themselves from predators, such as birds, lizards, bats, and spiders, by mimicry and camouflage.

After bees, butterflies are the second most common pollinators of crops.

Some butterflies are destructive. Cabbage white butterflies feed on cabbages and can destroy entire crops.

99

Tortoiseshell butterfly

Tortoiseshell butterflies are large, brightly colored butterflies found all over the world. They are one of the most common garden butterflies found in England.

Tortoiseshell butterflies can also be found in the dense hill forests of Asia and the northern United States.

Adult tortoiseshell butterflies feed on fruit juices or the nectar from flowers such as daisy and aster.

Female tortoiseshell butterflies are larger than the male butterflies.

After mating, a female tortoiseshell butterfly lays her eggs in batches on young nettle leaves. Each batch contains 60 to 100 eggs.

After about ten days, the eggs hatch and the caterpillars spin a web over the nettle's growing tip. Tortoiseshell caterpillars live in groups and feed on the nettle leaves.

Tortoiseshell caterpillars grow to about an inch in length within four weeks. Black, with two yellow broken lines along their sides, these caterpillars are poisonous.

Tortoiseshell butterfly pupae are grayish-brown and have metallic spots. These pupae are often seen hanging from posts, walls, and tree trunks.

Tortoiseshell butterflies used to be known as the "devil's butterfly" in Scotland.

▼ Tortoiseshell butterflies have a long life as adults, surviving for about ten months from one summer to another.

101

Red admiral butterfly

The red admiral butterfly is easy to recognize. It is black in color, with red bands and white markings on the upper and lower wings. Its wingspan is 1.7 to 3 inches.

Red admirals are found in gardens, orchards, and forests across Europe, North America, and Asia.

Adult red admiral butterflies feed on flowers and rotting fruits, while caterpillars feed on nettle plants.

This species is known for its migratory behavior. Red admiral butterflies cannot survive cold winters. Once the winter sets in, they migrate to warmer places.

The red admiral has a very erratic, rapid flight. Unusual for a butterfly, it sometimes flies at night.

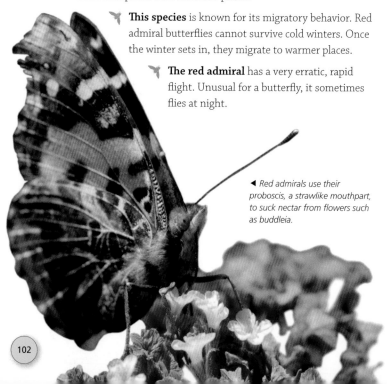

◄ Red admirals use their proboscis, a strawlike mouthpart, to suck nectar from flowers such as buddleia.

- **Female red admiral** butterflies lay their eggs on nettle leaves. After seven days, the caterpillar emerges and folds leaves around itself to make a protective tent. The leaves are held together with silk threads and the caterpillar feeds inside its leaf shelter.

- **The caterpillars** are black to greenish-gray in color and have a yellow line running along each side.

▲ Named after their "admirable" colors, red admiral butterflies are easily recognized by their dark-colored wings with red bands and white spots.

- **Adult red admiral butterflies** may hibernate in winter, storing enough fat in their bodies for survival.

- **Red admiral butterflies feed** on tree sap, rotting fruit, and bird droppings. They also feed on the nectar of plants such as common milkweed, red clover, aster, and alfalfa.

DID YOU KNOW?

Red admiral butterflies prefer moist woods, gardens, parks, marshes, and fields. During migration, these butterflies can be found in almost any habitat, from the tundra in Canada to subtropical countries near the equator.

103

Monarch butterfly

🦋 **Monarch butterflies are found** all over the world, except in cold regions. They are bright black and orange in color and have a wingspan ranging up to 4 inches.

🦋 **These butterflies use** their body color to frighten off enemies. The orange color is considered a warning sign.

🦋 **Monarch caterpillars feed** on milkweed plants, retaining the sap in their bodies even when they mature into butterflies. Birds attempting to eat monarchs dislike the taste and spit them out.

🦋 **Monarchs are beneficial** for crops, as they eat milkweed plants, which are weeds.

▶ Monarch caterpillars eat milkweed plants, which are poisonous to predatory birds.

▶ *Monarch butterflies travel up to 2,500 miles when they fly south for the winter.*

Monarch caterpillars are brightly colored, with bold black and white stripes.

Male monarch butterflies have dark spots called scent scales on their hind wings. Females do not have scent scales.

A monarch butterfly takes approximately one month to mature from an egg into an adult butterfly. Adult monarch butterflies feed on flower nectar and water.

Long-distance migratory insects, monarch butterflies guide themselves during migration using the position of the sun and the magnetic field of the earth.

Habitat destruction and changes caused by logging are constant threats to monarch butterflies. Spraying of pesticides for weed control kills milkweed plants. This endangers the habitat and food source of these butterflies.

Bhutan glory butterfly

Bhutan glory butterflies are found in Bhutan and the northeastern parts of Asia.

These butterflies prefer to live in grass fields and undisturbed forests.

The wings of Bhutan glory butterflies measure about 4 inches. They are black in color.

Bhutan glory butterflies breed twice a year from May to June, and then from August to October.

Very little is known about the life history of Bhutan glory butterflies.

Bhutan glory butterflies protect themselves from predators by absorbing poison from the plants they feed on.

Experts believe that these insects probably feed on the poisonous Indian birthwort plant.

These butterflies fly at altitudes of 5,500–10,000 feet in the mountains.

In the past, Bhutan glory butterflies were collected in large numbers. Now their numbers have been greatly reduced and they are very rare.

Today, Bhutan glory butterflies are listed as an endangered and protected species.

DID YOU KNOW?
When at rest, the Bhutan glory butterfly hides its colorful orange back wings with its front wings. This provides camouflage from predators.

▲ If disturbed, a Bhutan glory butterfly quickly opens and shuts it wings, exposing its bright orange markings. These sudden flashes of color may confuse a predator and give the butterfly time to escape.

Painted lady butterfly

▲ The painted lady butterfly can be distinguished from the tortoiseshell butterfly by the white marks on the black tips of its front wings.

DID YOU KNOW?
Painted lady butterflies do not survive in winter. They die in the cold weather.

- **The painted lady butterfly** is popularly known as the thistle butterfly because its caterpillar primarily feeds on thistle plants. It is also known as cosmopolite because it is found worldwide.

- **The painted lady** is one of the best-known butterflies in the world.

- **Painted lady butterflies** are mostly found in temperate regions across Asia, Europe, and North America, especially around flowery meadows and fields.

- **These butterflies are primarily** black, brown, and orange in color. They have a wingspan of 1.5–2 inches.

- **Female painted lady butterflies** lay eggs on plants such as thistles. After three to five days, they hatch into caterpillars. The caterpillars transform into pupae and finally emerge as colorful butterflies.

- **The caterpillar**, or larva, lives in a silky nest woven around the plant on which it feeds.

- **Adult painted lady butterflies** feed on nectar from flowers such as aster, cosmos, ironweed, and joe-pye weed.

- **An adult** painted lady butterfly lives for only two weeks.

- **Painted lady butterflies are strong fliers** and long-distance migrants. They can travel thousands of miles, sometimes with thousands of individuals flying together.

Birdwing butterfly

🦋 **Birdwing butterflies** belong to the swallowtail group of butterflies, which sometimes have tails on their hind wings, like the wings of a swallow.

🦋 **Male birdwing butterflies** are brightly colored with yellow, pale blue, and green markings.

🦋 **Female birdwing butterflies** have cream and chocolate-brown markings on their wings.

DID YOU KNOW?
Rajah Brooke's birdwing butterflies were named after the British Rajah Brooke of Sarawak by the famous naturalist Alfred Russell Wallace.

🦋 **Predators avoid** birdwing butterflies because they are poisonous and taste bad. Birdwing caterpillars feed on the pipevine plants and absorb poisons from them.

🦋 **A birdwing butterfly** lives for about seven months. These butterflies are listed as an endangered species. People are not allowed to hunt them.

◀ Queen Alexandra's birdwing butterflies were named in honour of the wife of King Edward VII of England.

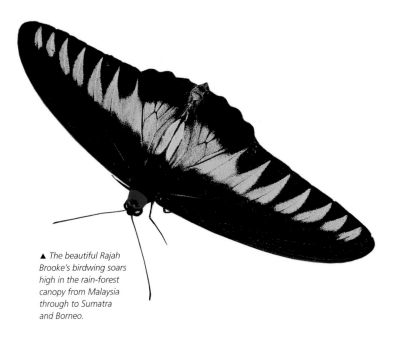

▲ *The beautiful Rajah Brooke's birdwing soars high in the rain-forest canopy from Malaysia through to Sumatra and Borneo.*

Birdwing butterflies are found in tropical areas. The best time to spot these butterflies is in the early morning, when they collect nectar from flowers.

Female Queen Alexandra's birdwing butterflies are the world's largest butterflies. They have a wingspan of up to 12 inches.

The golden birdwing and southern birdwing are some of the Queen Alexandra's birdwing butterfly's cousins.

Apollo butterfly

Apollo butterflies are mostly found in mountains and hilly regions of Spain, central Europe, southern Scandinavia, and Asia. Some Apollo butterflies live above an altitude of 13,000 feet and rarely descend to lower levels.

▼ *This mountain butterfly has a wingspan of 2–4 inches and a furry body to protect it from the cold. It survives the winter as a tiny caterpillar inside its egg.*

- **These butterflies are cream in color**, with red and yellow eyespots on their wings. They are frail-looking butterflies but are able to survive harsh weather conditions.

DID YOU KNOW?
An adult Apollo butterfly has a life span of just one week.

- **Habitat destruction** has made Apollo butterflies extremely rare. They are now an endangered species, protected by law in many countries.

- **The breeding season** of Apollo butterflies lasts from July to August. The female lays hundreds of eggs.

- **Female butterflies** lay round, white eggs either singly or in groups. The eggs usually hatch in August and September.

- **Apollo caterpillars** feed on stonecrop plants. These caterpillars molt five times in their lifetime.

- **Attempts are being made** to save Apollo butterfly populations by means of habitat management measures, reduction of insecticide use, and observation of their behavior during the time they are on in flight.

Peacock butterfly

- **Peacock butterflies** are named after their large, multicolored eyespots, which look like the "eyes" on a real peacock's feathers.

- **These butterflies inhabit** the temperate regions of Europe and Asia. They are very commonly found in lowland England and Wales.

- **Adult peacock butterflies** like orchards, gardens, and other places that have lots of flowers. They feed on the flower nectar of thistles, lavender, and buddleia and also suck juices from overripe fruits.

- **Peacock caterpillars feed** on nettle plants. They live in groups.

- **Peacock butterflies have** a single brood in a year. Adults hibernate through the winter and emerge in the spring. They die after laying eggs, while the caterpillars hatch out and hibernate as adults.

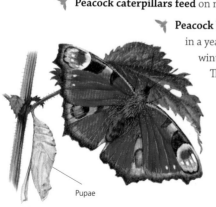

Pupae

◄ Female peacock butterflies often lay their small green eggs on nettles. Adults emerge from the pupae in July.

▶ Peacock butterflies
are often seen on
buddleia, attracted
by the tiny flowers'
sweet-tasting
nectar.

Female peacock butterflies can lay up to 500 eggs. The caterpillars emerge after one to two weeks and all live together in a communal web.

Fully grown caterpillars are about 2 inches long. They have black and white spots and long black dorsal spines.

The pupae are grayish-brown or greenish, with metallic gold spots.

Peacock butterflies have quite long life spans—they can live for as long as one year.

Viceroy butterfly

▲ *1. A Viceroy butterfly splits its chrysalis open by swallowing air, which makes its body expand. 2. It struggles free of its chrysalis. 3. The butterfly emerges and clings to the chrysalis. 4. It hangs from the chrysalis, pumping blood into its wing veins to stiffen and stretch them. 5. After about half an hour the wings are at their full size. Once the wings have dried, the butterfly is able to fly.*

Viceroy butterflies are mostly found in the United States, southern Canada, and northern Mexico.

These **butterflies have** black and orange patterns and white spots on their wings—resembling monarch butterflies.

They are found in meadows, marshes, swamps, and other wet areas with trees such as willow, aspen, and poplar.

There are usually two or three generations of viceroy butterflies born in each breeding season.

Viceroy butterflies are known to mate in the afternoon. The female butterfly lays her eggs on the tips of poplar and willow leaves.

The eggs hatch and the viceroy caterpillars feed on the leaves. They are voracious eaters and even eat their own shells. The caterpillars are white and olive-brown.

Adult viceroy butterflies feed on the liquids from decaying fungi, dung, and other animal waste.

Predators often mistake viceroy butterflies for monarch butterflies, and avoid eating them because monarch butterflies taste bad.

DID YOU KNOW?

Butterflies practice two kinds of mimicry–Batesian and Mullerian. In Batesian mimicry a harmless species of butterfly mimics a toxic species. In Mullerian mimicry two equally toxic species mimic each other for mutual benefit. Viceroy butterflies exhibit Batesian mimicry.

5

Moths

Moths make up about 90 percent of the insects that belong to the order Lepidoptera, which also includes butterflies. Insects that belong to this order have scaly wings.

Moths feed on nectar as well as other plant and animal juices. Some moths do not feed as adults because they do not have mouthparts.

▲ *Boldly patterned, magpie moths have black-and-white wings with yellow bands.*

In some species of moths, the females do not have wings.

Most moths are only active at night, but there are some species of moths that remain active during the day.

Like butterflies, some moths, such as hawk moths, migrate long distances.

Some moths have large spots on their wings. From a distance, these spots resemble the eyes of a fearsome animal and scare away potential predators.

Moths are masters of mimicry and camouflage.

Certain moths can be mistaken for bird droppings when they lie still on the ground. This helps them escape predatory birds.

The larvae of these moths are destructive in nature and feed on different types of natural fabrics, such as wool, cotton, linen, and even fur, feathers, and hair.

DID YOU KNOW?

Tiger moths produce high-pitched clicks at night to warn bats that they taste bad. Bats soon learn to link the clicks with the awful taste and avoid eating these moths.

▼ Unlike butterflies, which have clubbed antennae, some moths have feathery antennae.

Atlas moth

Atlas moths belong to the emperor moth group and are known for their large size. They are named after the patterns on their wings, which look like maps.

Atlas moths are the largest moths, with a wingspan of 9–12 inches. When they fly, these insects are often mistaken for birds.

◄ Giant atlas moths have transparent triangles in the middle of each wing where the colored scales are missing. These shiny patches may confuse predators by reflecting sunlight.

🦋 **These moths are found** in tropical forests and are natives of Southeast Asia.

🦋 **The tips of their wings** are hooked and have patterns on them, which helps to scare predators away. These marks resemble the head of a snake.

🦋 **Males** have large, feathery antennae. These antennae are capable of sensing pheromones released by female atlas moths from a distance of several miles. Female pheromones attract males for mating.

🦋 **Females are much larger** and heavier than the males. Their antennae are also less hairy.

🦋 **Adults** do not have mouthparts, so they cannot eat. They live for just two weeks and die soon after mating.

🦋 **Females lay eggs** under leaves. These eggs hatch into green caterpillars, which feed on leaves.

🦋 **Atlas moth caterpillars** have fleshy projections all over their bodies. These caterpillars can grow up to 5 inches in length.

Death's-head hawk moth

Death's-head hawk moths belong to the group of moths called sphinx moths. They are found in Africa, Asia, and Europe.

This moth is named after a peculiar mark that is visible on its thorax. The mark looks like a skull.

Death's-head hawk moths steal honey from bees' nests, which is why they are also known as "bee robbers."

At any point in its growth stage, a death's-head hawk moth is capable of producing a loud squeaking sound to scare its predators away.

▼ The caterpillars of death's head hawk moths reach 12.5 cm when fully grown and make a clicking sound if they are disturbed. They feed on potato plants and tomato leaves.

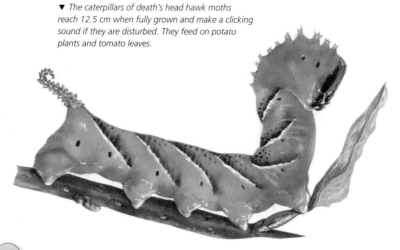

▶ The skull pattern on the moth's thorax includes eye sockets and a jaw, while the yellow bands on its body look like ribs.

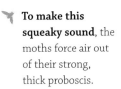

To make this squeaky sound, the moths force air out of their strong, thick proboscis.

Females lay single eggs on different plants. They prefer to lay their eggs on potato and brinjal plants.

The caterpillars have a horn on their tail end and are also known as hornworms.

To pupate, the caterpillars make a mud cell deep in the soil and smooth it by pushing their head against the wall of the cell.

Death's-head hawk moths find it difficult to survive harsh winters, and migrate to warmer places.

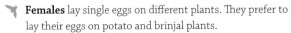

DID YOU KNOW?
The strange mark on the thorax of the death's-head hawk moth has given rise to many superstitions. In ancient times, the presence of this moth was considered to be a sign of death.

123

Gypsy moth

Gypsy moths are so called because they can travel long distances in a short time.

Female gypsy moths have cream-colored wings and cannot fly. Males are dark in color. Females produce chemical pheromones to attract males.

After mating, the female lays her eggs in clusters and covers them with body hair to protect them.

These eggs hatch into larvae, which feed on the leaves of different trees. They wriggle up a tree and spin silk threads. The larvae then swing from the silk thread and allow the wind to carry them away.

When the larvae land in a new feeding ground, they repeat this swinging activity. Older larvae do not swing from silk threads. They feed on leaves during the night and rest during the day.

The larvae spin a cocoon and pupate. Male larvae pupate earlier than female larvae and are smaller.

Gypsy moth larvae are considered to be pests because they feed on the leaves of various trees, such as oak, maple, and elm. They also feed on shrubs and other plants. Adult gypsy moths do not feed at all.

Trees that are affected by gypsy moths become leafless. Badly affected trees dry up and die.

Wasps and houseflies often attack gypsy moth larvae when they are resting on plants and trees. Birds and small animals, such as squirrels and mice, also eat these moths.

Gypsy moths are found in Europe, Asia, and the United States.

▼ A gypsy moth caterpillar. Gypsy moths were introduced into the United States in the 1860s, in order to develop a silkworm industry. The moths quickly established themselves in the wild and became a major pest.

Swallowtail moth

▼ *This large swallowtail moth has broad wings, like a butterfly, but flies rapidly. In June and July, it is widespread in Europe and parts of Asia, often flying around lights.*

- **Swallowtail moths** belong to the family Geometridae of the order Lepidoptera.

- **These moths are** strikingly unusual and can be mistaken for butterflies. They have slender bodies, thin legs, and a short proboscis.

DID YOU KNOW?

The swallowtail moth's name derives from a resemblance to swallowtail butterflies. Their hind wings have a tail similar to that of swallowtail butterflies.

- **Swallowtail moths are colorful** and sometimes fly in the day.

- **These moths** are mainly found in tropical countries.

- **Brilliantly colored species** are very large in size. Their colors are structural and do not contain pigments.

- **Some of the nocturnal species** have eyespots at the tips of their short, pointed hind wings.

  - **The eyespots** give an impression of a false head at the rear side of the moth, which protects it from predators. Therefore, during the day, the moths always rest on the upper side of leaves.

    - **The size and marks** on the bodies differ in males and females.

    - **Not much is known** about the life history of swallowtail moths.

Hummingbird hawk moth

- **Found all over the world**, hummingbird hawk moths belong to the hawk moth group.

- **Unlike other hawk moths**, these moths fly during the day and can be easily spotted hovering over flowers in gardens and parks.

- **Hummingbird hawk moths** are brownish in color. They have black and white spots all over their body and their hind wings are orange.

- **These moths** have tufts of hair at the tip of their abdomen.

- **Like other butterflies and moths**, hummingbird hawk moths have long, tubelike mouthparts that are coiled and tucked under their head. They use their long tongue to collect nectar from flowers.

- **Hummingbird hawk moth caterpillars** are slender and colorful. Their horned tail gives them a fearsome look.

- **The moth pupates** in leaf litter (dead leaves, bits of bark, and other dead plant matter lying on the ground) and weaves a very thick cocoon.

- **People often mistake** this moth for a hummingbird because it hovers over flowers and sucks nectar from them, like a hummingbird.

- **Hummingbird hawk moths** hibernate in winter to survive the cold weather.

▲ The broad body of this hummingbird hawk moth shows that it is a powerful flier. It holds its body still while hovering in front of flowers, beating its wings so fast that they are almost invisible. The rapidly beating wings produce a high-pitched hum, like the wings of a hummingbird.

Lobster moth

- **Lobster moths belong to** the group of moths known as prominent moths.

- **These moths** are commonly found in deciduous forests in Europe and Asia.

- **The wingspan** of lobster moths is about 2–3 inches.

- **Males** are often attracted to light, but females do not behave in the same way.

- **These moths** appear in two color forms—one with light front wings and one with dark front wings.

- **Lobster moths move** in a way that resembles the movements of ants.

- **The unusual shape** of lobster moth caterpillars often confuses their predators and scares them away.

- **As an act of defense**, the caterpillar curls back its large head and raises its legs in the air to startle small birds.

- **The caterpillar** constructs a silken cocoon and pupates in it.

> **DID YOU KNOW?**
> Lobster moths are so called because their larvae have six wiry, elongated legs and a swollen tail, like a lobster.

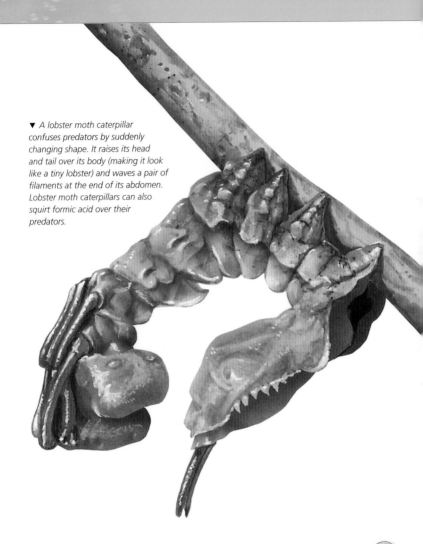

▼ A lobster moth caterpillar confuses predators by suddenly changing shape. It raises its head and tail over its body (making it look like a tiny lobster) and waves a pair of filaments at the end of its abdomen. Lobster moth caterpillars can also squirt formic acid over their predators.

Peppered moth

▶ The dark-colored form of the peppered moth is less common today in industrial areas where pollution controls have cleaned up the air. The dark color does not provide effective camouflage against clean tree trunks, which are often covered by pale-colored lichens.

Peppered moths belong to a group of moths known as geometrid moths.

These moths are delicate insects and have long legs and a slender body.

Peppered moth caterpillars do not have any legs in the middle of their body. They hold on to branches with their first two pairs of specialized limbs, called prolegs, and a clasper at their tail end.

Male peppered moths have feathery antennae, while females have hairlike antennae. The antennae of male peppered moths are also longer than those of females.

DID YOU KNOW?
Peppered moth caterpillars are known as inchworms because they move their bodies in a looping fashion as though they are measuring the earth.

🦋 **Males are smaller** and more slender in comparison to the larger and heavier female peppered moths.

🦋 **Peppered moths are nocturnal** and usually rest on lichen-covered trees during the day.

🦋 **Peppered moth caterpillars** camouflage themselves by resembling a twig.

🦋 **There are two varieties** of peppered moths: a paler-colored variety, speckled with salt and pepper (black-and-white) marks, and another that is coal-black in color.

🦋 **During the Industrial Revolution**, the bark of many trees became blackish-gray in color due to air pollution. Dark-colored peppered moths blended with this polluted environment and matched the color of the tree bark. Paler-colored peppered moths stood out as easy targets for birds and other hungry predators, so more dark-colored moths survived.

◀ The pale-colored variety of the peppered moth shows up better against dark backgrounds, making it more likely to be spotted by birds and other predators.

Moon moth

- **Moon moths belong** to the group known as emperor moths. They are found all over the world, but are mostly seen in tropical countries.

- **These moths** have a white body and maroon legs. Their wings are bright green in color. The color of female moon moths is brighter than that of the males.

- **Moon moths have a wingspan** of about 4–5 inches and their tail is almost 3 inches long.

- **The hind wings** of males are longer than those of the females. In overall size, male moon moths are much smaller than females.

- **Adult moon moths** have no mouthparts and do not eat anything. They do not live for more than a week.

- **Moon moths** make their cocoons in leaves. The silk of the cocoon does not shine and is brown in color. This silk is not used commercially.

- **An adult female** lays about 250 eggs at a time, on walnut leaves. The eggs resemble seeds with gray specks on them.

> **DID YOU KNOW?**
> Moon moths are also called "lunar moths" because of marks on their wings that look like a new moon.

- **Moon moths** grow only at the larval stage. If the larva grows into a small moth, the moth does not grow any more.

- **Moon moth caterpillars** are bright apple green in color and are beautifully segmented, with some white hairs.

▲ Moon moths have a long tail, which is mainly for show, to attract a partner. By day, as the moth rests, the tail looks like old leaves.

Horsefly

Horseflies are strong-bodied flies with colorful patterns on their bodies.

Female horseflies suck and feed on the blood of humans and animals. Different types of horseflies get their names, like deerflies or moose flies, depending on the animal they feed on.

There are about 25,000 species of horseflies in the world.

Horseflies have compound eyes, which are very prominent and occupy the entire surface of the head in the males.

▼ Horseflies are among the fastest flying insects, reaching maximum speeds of 24 mph. Unlike most other flies, their flight can be silent, allowing females to sneak up on their prey.

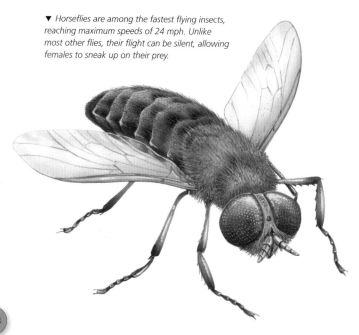

Colorful patterns on the compound eyes are caused by the refraction of light. There are no pigments present in the eyes and as a result, the color is not retained when these insects die.

The mouthparts of these insects consist of a short, powerful piercing organ capable of penetrating tough skin. If undisturbed, horseflies can suck blood from their host for as long as half an hour.

Female horseflies bite animals to suck blood before they reproduce. However, male horseflies do not bite. They feed mainly on flower nectar and plant sap.

After mating, females lay their eggs on plant and rock edges near water. The eggs are creamy white in color.

Horsefly larvae feed on soft-bodied insects and other small animals. They can become cannibalistic if there is a lack of food.

While biting and sucking blood from animals, female horseflies can transmit diseases such as anthrax.

▶ This horse is wearing a protective fly cover to prevent flies from biting it and possibly transmitting diseases.

Crane fly

🪰 **Crane flies** belong to the order Diptera and are closely related to mosquitoes.

🪰 **Their legs** are weakly attached to their body and often break off.

🪰 **Some of the larger species** of crane flies, such as phantom crane flies, have legs that are as long as an inch.

🪰 **Crane flies** have extremely narrow wings. They have a thin body with long legs and resemble large mosquitoes. These insects cannot bite.

🪰 **Crane flies are nocturnal insects** and remain inactive in the day.

🪰 **Females** lay their eggs in moist soil. The eggs hatch into larvae, which are grayish to pale brown in color.

🪰 **The larvae** of crane flies feed on dead and decaying matter. Some species also feed on small insects, and others eat plant roots.

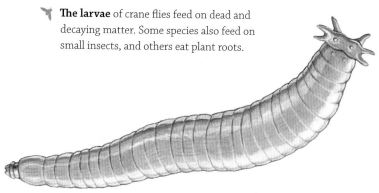

▲ *Crane fly larvae eat the roots of grasses, including cereal crops such as wheat.*

▲ *The long, thin legs of a crane fly are usually twice the length of its body. These insects can also be recognized by the V-shaped groove on the top of their thorax.*

Crane fly larvae are also known as leatherjackets because of their tough, brown skin. They are often used as fishing bait.

Farmers dislike the larvae of these insects because they damage the roots and turf of grain fields and grass crops.

A rare species of crane fly does not have wings. They are found in Hawaii.

Dobsonfly

- **Dobsonflies** belong to the order Neuroptera. They look like dragonflies.

- **Adults** have veins on their wings, which are gray in color. They are nocturnal insects.

- **Male dobsonflies** have mandibles (mouthparts) that are almost half the size of their bodies. These look like the long tusks of an elephant. Even though these insects look ferocious, they are quite harmless.

- **Females** have small, stout mandibles, which they use to defend themselves. They can inflict painful bites.

- **These mandibles** are not used for eating. In fact, adults do not eat anything. The male uses his mandibles to hold the female while mating.

- **After mating**, the female lays her eggs on tree branches and plants or on rocks near streams. The eggs look like bird droppings because they have a white coating.

- **The eggs hatch** and the larvae drop into the stream. The larvae are known as hellgrammites. They have gills and are dull in color.

- **Found under rocks** in streams, the larvae feed on other larvae and insects in the stream.

- **The larvae crawl** out of water and pupate near the shore. The pupae are found near the water, under logs, or in the soil.

- **Not usually** harmful to humans, the larvae only bite if disturbed. They are also used as fishing bait.

▶ *Female dobsonflies have much smaller mandibles than males, but their bite can be more painful.*

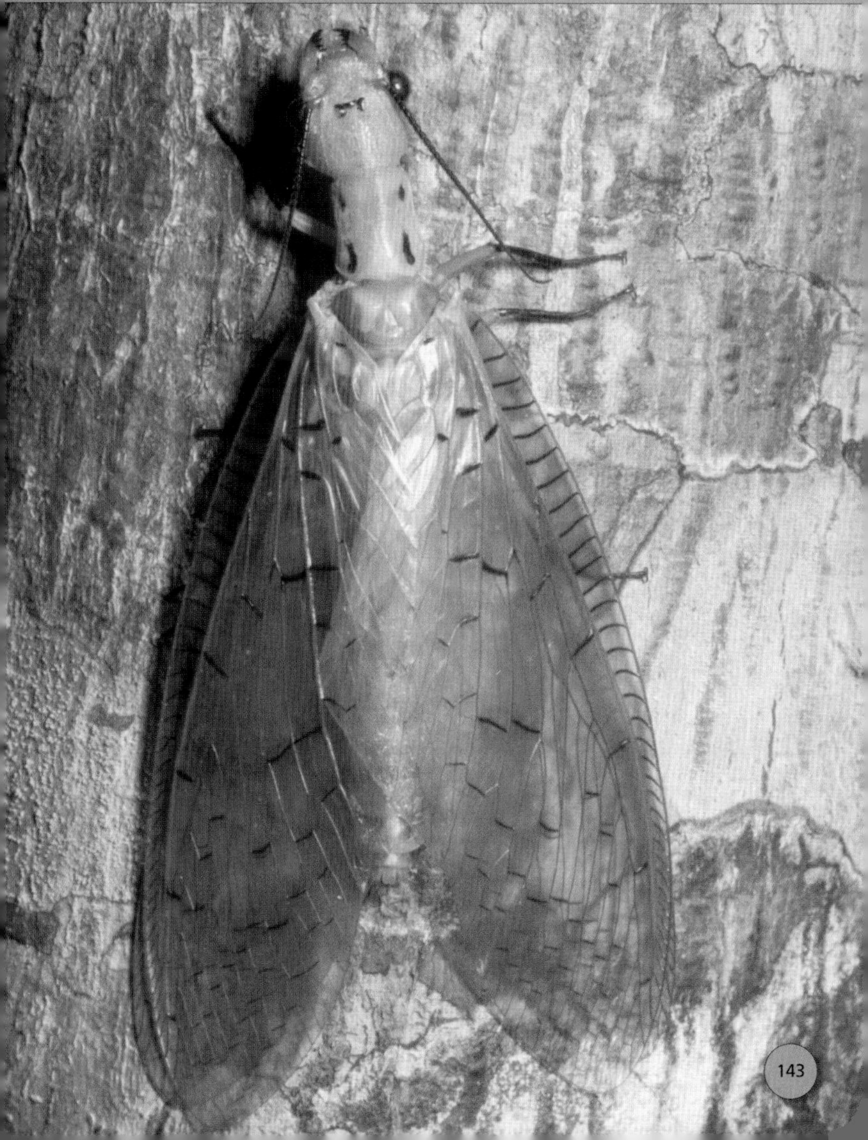

Housefly

Houseflies belong to the order Diptera and are one of the most common insects.

Like all flies, houseflies have small projections below their front wings known as halteres, which help them to fly. Halteres are modified hind wings, which are used for landing and balancing.

◀ Houseflies dribble saliva onto food. Powerful chemicals in the saliva break down the starch and other nutrients. They then dab up the liquid food with spongelike mouthparts.

▶ *Common houseflies are normally found wherever there is human activity. They thrive on the waste materials left behind by humans and other animals.*

Houseflies feed on liquid food and do not bite animals and other insects. The mouthparts of a housefly are like a sponge that absorbs liquid food.

Before eating, houseflies often vomit some portion of their last feed on top of new food particles. This makes the new food material easier to digest.

Most of the time, houseflies do not eat all the food and leave some particles behind. These remaining food particles can spread a variety of diseases.

Sticky pads and sharp claws enable houseflies to walk upside down with ease. Fine hairs on the tip of a housefly's legs enable it to "taste" liquids.

Houseflies have compound eyes and 4,000 individual lenses form each eye. These insects cannot see the color red.

Female houseflies can lay around 600 to 1,000 eggs in a lifetime. However, most offspring do not survive to reproduce.

DID YOU KNOW?

Houseflies are known to spread 40 serious diseases. A single fly harbors as many as 33 million infectious organisms inside its intestines and 500 million on its body surface and legs.

Robber fly

🦟 **Robber flies** belong to the family Asilidae. They are also known as assassin flies.

🦟 **These insects** are reddish-brown and have hair all over their body. Most species of robber flies have a long abdomen.

🦟 **Mostly found** in countries with a warm climate, these insects are normally active on warm and sunny days. Certain species of robber flies are also found in dense vegetation.

▲ *Giant robber flies are up to 1.5 inches in length and are found in North America.*

🦟 **Some robber flies** mimic bees to escape from predators. They are stouter than bees and their abdomen is not as elongated.

🦟 **Predatory insects**, robber flies are fast fliers. They hunt while flying and can attack insects that are much larger and stronger than they are.

🦟 **Robber flies attack insects** such as butterflies, dragonflies, bees, and wasps. They hold their prey with their front limbs and then inject saliva into it. This saliva paralyzes and partly digests the victim.

DID YOU KNOW?
After mating, the female robber fly scatters her eggs on the ground. Some species also lay eggs in plants. These eggs are long in shape, while the eggs laid on the ground are round.

🦟 **The body fluids** of the prey are sucked out. Robber flies can only digest fluids, so their mouthparts are suitable for piercing and for sucking liquids.

🦟 **Male robber flies** sometimes perform a dance ritual before mating, which includes hovering in front of the female.

🦟 **Eggs hatch** into larvae that live in soil and rotting wood. Some experts say that robber fly larvae are herbivores, while some suggest these larvae also feed on other insects.

▼ Huge compound eyes provide the robber fly with the excellent vision it needs to be an effective predator. Long bristles on the face help to protect the robber fly's face when it attacks its prey.

Mosquito

Mosquitoes are the only flies of the order Diptera that have scaly wings.

Most female mosquitoes have to feed on the blood of other animals to reproduce. They need the protein extracted from blood for the development of their eggs.

These eggs float on water and hatch to produce aquatic larvae known as wrigglers.

The larvae cannot breathe underwater and "hang" from the water's surface to take in air.

Mosquito larvae pupate in water. The pupa is not completely immobile. It can change position according to the light and wind conditions in its environment.

Adult male mosquitoes feed on nectar and other plant fluids. Only female mosquitoes feed on blood.

🦟 **Male mosquitoes** cannot bite. Their mouthparts are modified for sucking only.

🦟 **Mosquitoes** have infrared vision. They can sense the warmth of other insects and animals.

🦟 **Dragonflies feed** on mosquitoes and are also known as mosquito hawks. Dragonfly nymphs also feed on mosquito larvae.

🦟 **Mosquitoes are known** to spread many infectious diseases, such as yellow fever and dengue fever. The female anopheles mosquito spreads malaria.

◀ *Mosquitoes have long mouthparts that they use to pierce skin (human or animal) and suck up blood.*

Firefly

🦟 **Fireflies are not flies** but a type of beetle. These insects are also known as lightning bugs.

🦟 **These insects can grow** up to an inch in length and live for three to four months. Females live longer than males.

🦟 **Fireflies** have a special organ under their stomach, which emits flashes of green or yellow light.

🦟 **This light** is known as "cold light." The process of producing light is also known as bioluminescence. No other insect can produce light from their own bodies.

🦟 **Fireflies emit** only light energy. A normal lightbulb emits only 10 percent of its energy as light. The remaining 90 percent is emitted as heat energy. Fireflies do not emit any heat energy.

◀ *A pair of fireflies mating on a leaf.*

- **Males and females** emit light to attract mates.

- **Fireflies are nocturnal** insects. They live on plants and trees during the day and are active at night.

- **Females** and their larvae are carnivorous and feed on snails, slugs, and worms. Most males don't eat, but some may feed on pollen and nectar.

- **Some females are wingless** and are also known as glowworms.

DID YOU KNOW?
Some female fireflies fool males of other species by flickering their light. When the males come closer to mate, the females eat them.

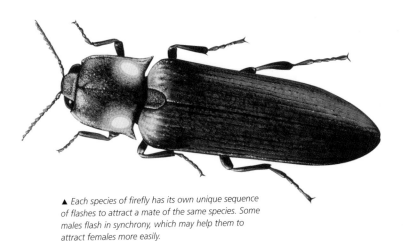

▲ Each species of firefly has its own unique sequence of flashes to attract a mate of the same species. Some males flash in synchrony, which may help them to attract females more easily.

151

Goliath beetle

One of the largest and heaviest insects in the world is the Goliath beetle. Males are as heavy as an apple.

Goliath beetles grow up to 4.7 inches and weigh about 4 ounces.

Found in many colors, most Goliath beetles have black and white markings on their wings.

▲ *Goliath beetles are so big, they make a whirring noise like a model plane when they fly.*

Goliath beetles used to be found only in Africa. Today, however, these insects can be found in almost all parts of the world.

Males have a horn-shaped structure on their heads. They often fight each other with their horns.

Goliath beetles are good fliers. They produce a low, helicopter-like whir while flying.

By feeding on dead plant and animal tissues, Goliath beetles help to keep the environment clean.

A fossil of the oldest Goliath beetle is almost 300 million years old.

Scientists believe that Goliath beetles, like some other insects, were much larger in prehistoric times. This may be because there was more oxygen in the air then.

DID YOU KNOW?
The Goliath beetle is named after the biblical giant Goliath.

Diving beetle

- **Diving beetles** have oval bodies, which are brown or black in color. Their wings are shiny and look metallic green when light reflects off the body. These beetles are commonly found in ponds and other still water.

- **A diving beetle** has strong, hairy legs, which it uses like the oars of a rowboat to push its body forward when swimming.

- **These insects** cannot breathe underwater and have to come to the surface to breathe. They store air under their wing covers.

- **Diving beetles often fly** at night in search of new ponds.

- **These beetles locate** new ponds with the help of the light that reflects from water surfaces. They often get confused with light that is reflected from a glass surface and mistake it for water.

🪲 **The female diving beetle** is much bigger than the male.

🪲 **Females** make small slits in plant stems and lay their eggs inside.

🪲 **The larvae float** on the surface of the pond and move to the shore to pupate.

🪲 **The larvae** are known as "water tigers" because they can attack and bite other insects. They even eat tadpoles and small fish.

◄ *Diving beetles feed on other water insects, small fish, and tadpoles.*

155

Tiger beetle

Tiger beetles are usually shiny metallic colors, such as green, brown, black, and purple. They often have stripes like those of tigers.

The smallest tiger beetles live in Borneo and measure up to 0.2 inch. The largest tiger beetles live in Africa and can reach up to 1.7 inches in length.

Tiger beetles are found in sunny, sandy areas and are active during the day. However, there are some species that come out at night.

Tiger beetles are good fliers and fly in a zigzag pattern if a predator approaches them. They are also very fast runners.

▼ *Tiger beetles have huge eyes. They use their massive biting jaws to catch and cut up their food.*

▶ Tiger beetles are often very colorful, with distinct markings.

DID YOU KNOW?
In Borneo, a type of locust mimics tiger beetles to protect itself from its predators.

🦟 **These insects are predatory**. Once they locate their prey, they pounce on it and use their jaws to tear it into pieces.

🦟 **Tiger beetles** taste awful and predators avoid eating them.

🦟 **Females** lay their eggs in burrows in the ground. The burrow can be almost 20 inches deep.

🦟 **The larvae** have a pair of powerful and large jaws, which are used to capture small insects.

🦟 **Some species** of tiger beetles are endangered or nearing extinction. This is because of the lack of undisturbed sandy areas in which they can breed.

Stag beetle

Stag beetles are usually brown or black in color, but there are some species that are bright green and red.

These beetles live in damp wooded areas, especially near oak forests.

Stag beetles have mandibles (jaws) that resemble the antlers of a stag.

Males have long and ferocious-looking jaws. The jaws of females are not as long as those of the males.

▼ *Stag beetles belong to the scarab beetle family, which contains over 20,000 species.*

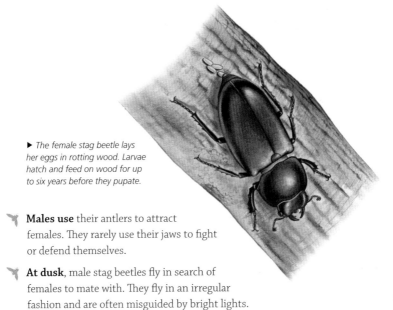

▶ The female stag beetle lays her eggs in rotting wood. Larvae hatch and feed on wood for up to six years before they pupate.

Males use their antlers to attract females. They rarely use their jaws to fight or defend themselves.

At dusk, male stag beetles fly in search of females to mate with. They fly in an irregular fashion and are often misguided by bright lights.

The larvae of these insects feed on rotten wood and plant remains.

Larvae take a long time to develop into mature adults because the food on which they feed is not nutritious.

Adult stag beetles do not usually eat anything because they have enough reserve energy stored as fat. However, they sometimes feed on the sweet sap of trees.

DID YOU KNOW?
Stag beetles are also known as "pinching bugs" because they can nip with their jaws and draw blood from human beings.

159

Rhinoceros beetle

🐾 **Rhinoceros beetles** are believed to be the strongest creatures on earth. They can carry about 850 times their own weight.

🐾 **These beetles can grow** up to 5 inches in length.

🐾 **The preferred habitat** of rhinoceros beetles is tropical rain forests, where the vegetation is thick and there is plenty of moisture in the atmosphere.

🐾 **Rhinoceros beetles** are named after the horns on their head, which resemble spikes.

🐾 **Only males** have horns. The larger the horn, the better their chances of winning a mate's attention.

🐾 **The horns** are very strong and can pierce through the exoskeletons of insects.

🐾 **Rhinoceros beetles do not use** their horns for defending themselves from predators. Instead, they use them to fight with other males for food and to attract female beetles for mating.

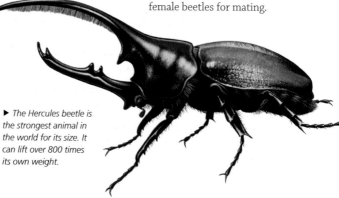

▶ The Hercules beetle is the strongest animal in the world for its size. It can lift over 800 times its own weight.

▲ This rhinoceros beetle has three horns, but others have two or five horns. The very tough exoskeleton protects the beetle's body like a suit of armor.

Rhinoceros beetles are nocturnal in nature and almost all their hunting and feeding activities take place at night.

These beetles help to keep the jungle clean and feed on plant sap and rotten fruits that have fallen to the ground.

Rhinoceros beetles are known to be fierce fighters. Natives in some parts of Thailand hold beetle fighting competitions to watch these beetles fight.

161

Ladybug

Ladybugs were dedicated to the Virgin Mary. They were known as the "beetle of Our Lady" during the medieval period.

Ladybugs are found in temperate and tropical regions all over the world.

These beetles are one of the most beneficial insects because they feed on insect pests that damage crops.

▼ *The number of spots on a ladybug differs from species to species. The two-spot ladybug is smaller than its seven-spotted cousin.*

Ladybugs are sometimes bred on a large scale and are then introduced into farms or greenhouses to get rid of pests. However, some species of ladybugs are herbivorous and are considered pests themselves.

These insects are easily mistaken for leaf beetles because of their similar coloring and spots.

▲ Brightly colored beetles, ladybugs have round bodies and hard wing cases, called elytra.

When disturbed, ladybugs secrete a foul-smelling fluid that causes a stain. This is called reflex bleeding.

Both adults and larvae eat aphids, scale insects, and other soft-bodied insects.

Ladybug larvae do not have wings. They are metallic blue in color, with bright yellow spots all over their body. Their bright color warns birds not to eat them.

The Halloween ladybug is a pumpkin-orange beetle found in the United States during late October.

Adult ladybugs hibernate in huge clusters in densely vegetated areas, usually at high altitudes.

True bugs

- **True bugs** belong to the suborder Heteroptera of the order Hemiptera. There are at least 55,000 different kinds of bugs.

- **These bugs** generally have two pairs of wings. The first pair is partially hard and protects the delicate, membranelike second pair of wings.

- **Some bugs** do not have wings, while the nymphs of all true bugs are wingless.

- **True bugs have compound eyes**, and their mouthparts are adapted for sucking and piercing. Most bugs have beaks that are segmented into four or five parts.

- **Bugs undergo incomplete metamorphosis**. There is no pupal stage and the bugs grow into adults by molting again and again. The nymphs resemble the adult bugs.

- **Bugs can survive** on land, in air, on the surface of water, and even underwater. There are very few places where bugs do not live.

- **Some bugs can give out** a very bad odor. This is a defense strategy.

Bugs feed on plant and animal juices. There are some bugs, such as bedbugs, that are parasites. They live by sucking blood from other animals.

Carnivorous bugs are predatory and help to control pests, while herbivorous bugs are a threat to crops. Bugs can be cannibalistic and may feed on weaker individuals of their own kind.

▼ *The word "bug" is sometimes used to mean any insect, but a true bug is an insect with piercing and sucking mouthparts, which are tucked beneath the head when not in use.*

Assassin bug

Assassin bugs are either black, brown, or bright red and black in color.

The wings of assassin bugs lie flat on their abdomen. These insects have long legs, which are adapted for running.

▼ *Assassin bugs give out a pungent smell, which, along with their poisonous bite, protects them from their predators.*

Assassin bugs are predatory. They grab their prey and "assassinate" it by injecting venom. The venom paralyzes the prey, and partially dissolves and disintegrates it. The bug then sucks the liquid food.

DID YOU KNOW?

The saliva of assassin bugs can cause temporary blindness in humans.

This venom is so powerful that caterpillars, which are much larger than assassin bugs, can be killed in seconds. An assassin bug may take several days to eat a large victim. Its front legs have powerful muscles for holding prey while sucking out their body fluid.

Assassin bugs have a powerful curved beak, which is used for sucking the blood of insects, larger animals, and even humans.

Male and female assassin bugs are similar in appearance but sometimes the females do not have wings.

An assassin bug's bite is quite painful and can transmit germs and diseases. These bugs are known to spread a disease called Chagas' disease.

A species of assassin bug known as the masked assassin bug camouflages itself by sticking dirt to its body.

Squash bug

- **Squash bugs are so named** because they are a threat to squash and other related plants. They can usually be found in colors ranging from brown to black.

- **Some squash bugs** have leaflike extensions on their hind legs. This makes them look like dead leaves, which helps with camouflage. These bugs are called leaf-footed bugs.

- **With their powerful beaks,** squash bugs can easily pierce and suck fluids from plants and insects.

- **Squash bugs have scent glands** that emit a pungent smell. However, the odor is not as strong as that of stinkbugs.

- **Although most** squash bugs are carnivorous, a few feed on both plants and insects, and some are strictly herbivorous.

- **While feeding**, the squash bug injects a toxic substance into the plant. As a result, the plant wilts and dies.

- **Squash bugs lay their eggs** in clusters on plants. The eggs are oval, flattish, or elongated in shape.

- **The nymphs** that hatch from these eggs resemble black ants and molt four or five times before maturing into adults.

- **Farmers consider** squash bugs to be dangerous pests and adopt various measures to get rid of them.

DID YOU KNOW?

A rice field affected by squash bugs can be smelled from a considerable distance.

▼ *The squash bug of North America feeds on the juices of cucumber, squash, melon, pumpkin, and other gourds. It does not have leaflike back legs.*

Cicada

Cicadas belong to the order Homoptera and are related to true bugs. Most species of cicadas are found in deserts, grasslands, and forests.

Cicadas have large and colorful wings. They hold their wings in a slanting position over their abdomen, like a tent.

Male cicadas emit sounds similar to those of a knife grinder, a railway whistle, and even oil spitting in an overheated pan.

Male cicadas sing loudly to attract females, with the help of special drumlike membranes called timbals. Female cicadas do not produce any sound because their timbals are not developed.

Large swarms of cicadas attract birds, which feast on these insects.

Female cicadas lay their eggs on plants and tree twigs. When the eggs hatch, the nymphs fall to the ground. They live underground for many years, feeding on the roots of plants. Later, they emerge from the ground, climb up trees, and then molt.

Adult cicadas do not live as long as the nymphs. They only survive a few weeks.

A species of cicada known as the periodical cicada is found in North America. It emerges from under the ground every 13 or 17 years. These insects are one of the longest-lived in the world.

▶ *Cicada eggs hatch into wingless nymphs. Most cicadas spend between one to three years as nymphs, growing and molting as they gradually develop into adults.*

Spittlebug

▶ *Spittlebug nymphs produce "cuckoo spit" by giving off a sticky liquid and blowing it into a frothy mass of white bubbles.*

Spittlebugs are different shades of yellow and brown. They have a triangular head, red eyes, and spotted wings.

Also known as froghoppers, spittlebugs have a froglike appearance. They are good jumpers but rarely fly, even though they have well-developed wings.

Adult spittlebugs have rather large heads in comparison to their small bodies.

Females lay their eggs in the stems of grasses and other plants.

These bugs are named after the spitlike frothy mass secreted by the nymphs. This froth is sometimes known as "cuckoo spit."

DID YOU KNOW?
In Madagascar, spittlebugs discharge a clear liquid instead of foam, which falls to the ground like rain.

Spittlebug nymphs are not easy to spot because they are often hidden in cuckoo spit on leaves.

Cuckoo spit helps the nymphs to control their temperature and even prevents them from losing moisture and drying out.

▶ Cuckoo spit protects the spittlebug nymphs from predators.

Some spittlebug nymphs form delicate tubes about 0.4 inch in length. They attach these tubes along the sides of twigs and live there after filling them with spittle.

Spittlebugs are considered to be pests, as they feed on the sap of plants.

▼ Adult spittlebugs do not produce foamy spittle like their nymphs. They spend their time hopping about like tiny frogs on the plants and shrubs from which they feed.

173

Mealybug

Mealybugs are very small insects, up to 3 mm long. They are found in huge clusters on leaves, twigs, and tree bark. These bugs are also called coccids.

Females do not usually have wings, eyes, or legs and remain immobile on plants. They are always covered in a white sticky coating of their own secretion. This protective coating looks like cornmeal.

Males usually have one pair of delicate wings, well-developed legs and antennae, and no mouthparts.

▲ Female mealybugs secrete a mass of waxy threads as a protective covering. They are often wingless and legless, with reduced antennae.

▶ *Mealybug nymphs do not usually move after their first molt. They stay in one place, joined to plants only by their sucking mouthparts.*

Female mealybugs never lay their eggs in the open. The eggs are attached to their bodies. However, some species of mealybugs can give birth to live young.

The flat, oval larvae crawl about quite actively at first, but soon lose their legs and cover themselves with their mealy secretions.

A male mealybug's hind wings are modified into tiny structures called halteres.

Mealybugs feed on the sap extracted from plant tissues and are considered to be pests of citrus trees and greenhouse plants.

These bugs can harm plants in many ways. Galls can form on plants, or their stems can become twisted and deformed.

Mealybugs produce a sugary substance known as honeydew. Ants often visit these insects for this sweet-tasting secretion.

Beetles, lacewings, and caterpillars prey on mealybugs.

Aphid

Aphids are related to cicadas. They can be green, red, or brown in color and are also called greenflies.

These bugs have a large, pear-shaped abdomen with two slender tubes called cornicles attached to it. The cornicles secrete wax.

Aphids may or may not have wings. Those with wings are weak fliers, but they can cover great distances with the help of air currents.

Many aphids live underground and suck sap from roots. Sometimes these insects depend on ants to carry them through tunnels in the soil and leave them out on fresh roots.

▼ *Aphids produce a sugary substance that ants like to eat. In return, the ants protect the aphids from their predators, build shelters, "graze" them in fresh pastures, and even take them into their ant nests during bad weather.*

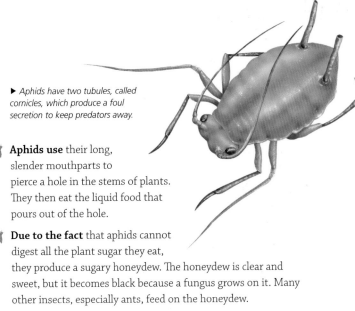

▶ Aphids have two tubules, called cornicles, which produce a foul secretion to keep predators away.

Aphids use their long, slender mouthparts to pierce a hole in the stems of plants. They then eat the liquid food that pours out of the hole.

Due to the fact that aphids cannot digest all the plant sugar they eat, they produce a sugary honeydew. The honeydew is clear and sweet, but it becomes black because a fungus grows on it. Many other insects, especially ants, feed on the honeydew.

Some aphids lay eggs, but others can reproduce without mating. These aphids give birth to live young by the process of parthenogenesis.

Aphids breed in huge numbers. A single aphid can produce up to 100 young at a time.

Some aphids secrete long, curly strands of a waxlike substance from their cornicles. These aphids form dense colonies, which can be seen easily on plants.

Aphids can cause damage to plants. Their saliva causes plant leaves to fold, curl, and even form galls.

177

Leafhopper

 Related to cicadas, leafhoppers have a distinct leaflike shape and color, and are easy to recognize.

 Leafhoppers can survive in almost any part of the world. They are terrestrial bugs and can be found in deserts as well as in marshy and moist places.

 Strong fliers, leafhoppers are also capable of jumping considerable distances.

 If these insects feed on plants, they can cause the leaves to curl and affect the plant's growth.

 Leafhoppers are considered pests because they damage crops.

 Leafhoppers search for mates by making special mating calls.

 Females lay their eggs in slits made in plant stems. The eggs can remain dormant for a month. In some species, the eggs remain dormant for a year.

 Like aphids, leafhoppers produce honeydew from the excess plant sap that they feed on.

 These bugs communicate with each other by producing low-frequency sounds, which humans cannot hear.

 Birds, reptiles, and large insects, such as wasps, are a threat to leafhoppers.

▲ *Leafhoppers are common jumping insects that look like a narrow version of an adult spittlebug. They are tiny insects—only 2–15 mm long.*

Stinkbug

🦟 **Stinkbugs have glands** on their undersurface, from which they secrete a fluid. This fluid has a foul odor, which is how these insects got their name.

🦟 **Their scent** has a strong effect on many animals and protects them from predators.

🦟 **Stinkbugs can be found** in many colors, such as green, gray, brown, red, black, and yellow. They are one of the most notorious pests found in farmlands and orchards.

🦟 **Some species** of stinkbugs produce a loud sound to defend themselves.

🦟 **Stinkbugs have a compound** called glycerol in their blood. This prevents their blood from freezing in winter.

🦟 **Many species** of stinkbugs are active during the day, while others are nocturnal, especially the dark-colored species, which live in thick grass or under leaves.

Stinkbugs normally feed on the sap from flowers, leaves, and fruits, which is why they destroy crops.

Male and female stinkbugs are very similar in appearance even though the male is much smaller in size.

Young nymphs do not have wings, but adult stinkbugs are very good at flying.

In some parts of Mexico, India, and Africa, people eat stinkbugs.

◀ In some stinkbugs, the thorax extends down the back to form a protective shield, which almost covers the abdomen. These stinkbugs are also called shield bugs.

181

Lesser water boatman

▲ Lesser water boatmen are not buoyant enough to float, so when they stop swimming, they sink to the bottom. This is useful because they feed on the bottoms of ponds, canals, and ditches, using their shovel-like front legs to dig up food.

Lesser water boatmen are aquatic insects and are usually found in ponds and other freshwater bodies, such as canals and ditches.

These bugs have powerful, hairy legs and are named after their habit of using their legs as oars to "row" themselves through the water.

DID YOU KNOW?

People in Mexico dry lesser water boatman eggs and eat them. In some countries, lesser water boatmen are sold as bird food.

Most species of lesser water boatmen can fly. However, these insects usually cling to water plants or live at the bottoms of ponds.

Lesser water boatmen do not have gills and have to come to the water's surface to breathe.

Hairs on the body trap air bubbles. This helps lesser water boatmen to stay underwater for long periods.

Lesser water boatmen feed on algae, plants, and decaying animal matter.

Some males rub their legs together to make squeaky sounds that attract females.

The female lays her eggs underwater. The young nymphs that hatch from the eggs resemble the adults (without wings), but they molt five times before they become winged adults.

Backswimmer

🦟 **Backswimmers** spend almost their entire life floating in an upside-down position in ponds.

🦟 **These insects use their legs like oars** and resemble rowboats.

🦟 **While floating**, backswimmers navigate with the help of light.

🦟 **Unlike other insects**, the underside of backswimmers is dark in color while the top of the body is light. The dark color helps to hide the backswimmer from predators.

🦟 **Backswimmers** come to the water's surface for air. They carry a bubble of air under their belly, which helps them to stay underwater for a long time. This air bubble gives them a silvery appearance.

🦟 **In appearance**, backswimmers look very similar to lesser water boatmen. However, lesser water boatmen float the right way up and feed mainly on plants. Backswimmers, in contrast, are predators.

DID YOU KNOW?

Backswimmers, which are also known as water bees, have a painful bite. They have been known to bite humans.

🦟 **These insects** fly from one pond to another in search of food.

▲ *Backswimmers spend almost all their time floating upside down.*

Backswimmers feed on other water insects, worms, and tadpoles.

When attacking, backswimmers inject toxins into their prey. This has a chemical reaction, which kills the prey.

Some species of backswimmers are known to hibernate. During winter, these insects are found moving under the surface of frozen water.

Water strider

Water striders are often found floating on still water and slow-moving streams. They are one of the few insects that can survive in the sea and have even been found floating on the surface of the Pacific Ocean.

▼ The long, thin legs of a water strider help to spread out its weight. The surface of the water bends into small dips around the end of each leg, but does not break.

- **Water striders** seldom go underwater. Their long legs help them to skate and steer across the water. They are also called pond skaters.

- **These insects have six legs**, although it seems as if they have only four. The front two legs are short and are mainly used to grasp prey. The hind legs are very long and can be twice the length of the body.

DID YOU KNOW?

Water striders are also known as "Jesus bugs" because of their ability to walk on water. These insects move very quickly and can run across water. If they were as big as humans, they would move as fast as a jet plane.

- **Water striders have** long, slender, hairy legs, which end in a pad of water-repellent hairs so they do not break through the surface of the water.

- **Some adults** have wings, while others are wingless.

- **Water striders feed** on insects that fall into the water and aquatic insects that come to the surface to breathe.

- **Water striders communicate** with each other by making vibrations and ripples on the water's surface.

- **Adults** hibernate throughout the winter and mate in the spring.

Water scorpion

Water scorpions look very similar to land scorpions. They have forelegs that resemble pincers. The abdomen is also long, like a scorpion's.

There are two types of water scorpions. One looks like a leaf, while the other is long and slender like a needle and is called a needle bug. Water scorpions are also called toebiters because they can sting people who are wading in the water.

Water scorpions live in plants at the water's edge. They move slowly and often stay still for many hours.

These insects are voracious eaters and feed on tadpoles, fish, and other small aquatic animals. Even though they can swim for short distances, they usually keep still and lie in wait for their prey. When the prey comes near, they pounce on it.

The prey is injected with venom and paralyzed. Water scorpions are true bugs and can only suck fluids out of their prey.

Like most aquatic insects, water scorpions do not have gills and their body hair traps bubbles. These air bubbles help them to breathe underwater.

Water scorpions occasionally come to the surface to breathe. They have a breathing tube attached to the end of their body that is used as a snorkel. However, the larvae breathe through spiracles (breathing holes in the side of the body).

Male water scorpions rub their forelegs against their body to make chirping sounds in order to attract a female.

🗡 **Females lay their eggs** in slits made in plant stems. These eggs have tiny holes, which help them to breathe.

🗡 **Water scorpions** are active in summer, but they survive winter as well. They are found floating in ponds even when the temperature is very low.

▼ *The water scorpion is a predator. Its legs are suited to catching prey, such as this tadpole, and its mouthparts are adapted to sucking fluids from the prey's body.*

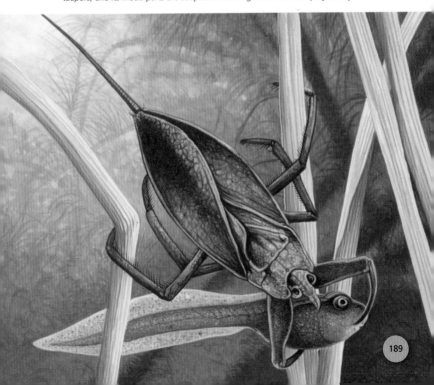

189

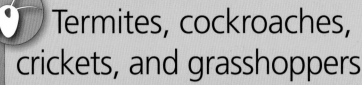
Termites, cockroaches, crickets, and grasshoppers

Termite anatomy

Termites are believed to have evolved from a cockroachlike ancestor at least 250 million years ago.

Termites live mainly in tropical places such as Australia, but some do live in cooler places in North America and Europe.

These insects are popularly known as white ants because termites and ants look similar, even though they are not related. However, they can be easily distinguished. Ants have elbowed antennae, while termites have straight antennae.

Termites also have two pairs of large wings, which overlap each other on their backs when at rest. These wings are equal in size, unlike those of ants. Termites belong to the insect order Isoptera, meaning "equal wings."

Leaving behind a stublike wing base, termites often shed their wings. Winged termites are called alates. They hatch once a year.

Termites are small and pale and avoid sunlight. Like all insects, they have a head, thorax, and abdomen. The thorax has a broad joint with the abdomen, unlike the pointed "waist" of ants, bees, and butterflies.

▼ *Initially, the queen lays a small number of eggs. As the colony matures, she can lay more eggs—as many as 36,000 eggs a day. The queen may grow to such a size that she can no longer move.*

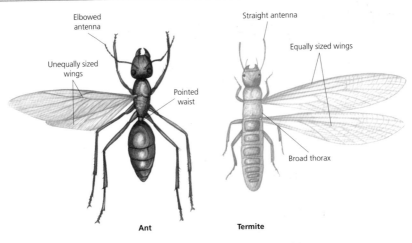

Elbowed antenna

Unequally sized wings

Pointed waist

Straight antenna

Equally sized wings

Broad thorax

Ant

Termite

▲ *A comparison of an ant and a termite. Although these insects are often mixed up, there are some key differences in their bodies.*

There are two main types of termites—ground termites (also called subterranean termites) and wood termites. Ground termites nest in soil and wood termites nest in wooden planks and other wooden structures.

The mouthparts of termites are modified for the purpose of chewing wood. They have antennae, which can be either beadlike or threadlike.

Termites have a very soft cuticle (an outer covering), which dries up easily. This is why they live in dark, warm, and damp nests.

Some termites have protozoa (single-celled creatures) in their stomach. These help with the difficult process of digesting cellulose (a carbohydrate that contains glucose), which is present in plants and wood.

193

Social behavior

🐦 **Termites** are social insects that live in well-organized groups or colonies. A termite colony is a highly integrated unit where each and every function is divided among the various termites.

🐦 **Small termite** colonies can have hundreds or thousands of termites, while a large colony can be home to millions of termites.

▼ *Termite mounds are incredibly complex constructions. They can reach 30 feet tall and have air-conditioning shafts built into them. These enable the termites to control the temperature of the nest to within one degree.*

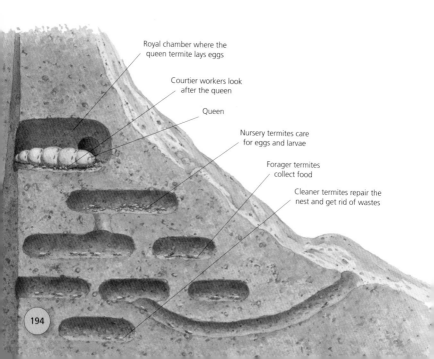

Royal chamber where the queen termite lays eggs

Courtier workers look after the queen

Queen

Nursery termites care for eggs and larvae

Forager termites collect food

Cleaner termites repair the nest and get rid of wastes

- **There are different categories** of termites in a colony—the king, the queen, workers, and soldiers. All these categories are referred to as castes. The workers and soldiers cannot reproduce.

- **Workers form** the major part of the colony's population. They work hard, taking care of the eggs and maintaining the nest, as well as foraging for food and feeding the young.

- **Soldiers defend the colony** from predators. Two large jaws, which protrude from a hardened head, are used as weapons.

- **In a colony**, termites communicate with each other about the direction and presence of food through vibrations, physical contact, and chemical scents called pheromones.

- **Reproductives** and soldiers produce a pheromone that helps them to maintain the caste balance of the colony. They lick the nymphs and transmit the pheromone that prevents these nymphs from developing into reproductives and soldiers.

- **If the number** of these termites goes down, they stop passing on the pheromone. This means that more reproductives and soldiers are produced, which restores the balance of the colony.

- **The king termite** helps the queen to set up the colony and mates with her. The queen lays eggs and looks after the colony.

- **Workers and soldiers** are the first nymphs to hatch from eggs so that they can build the nest and defend it. Once there are enough worker termites, the queen no longer looks after the young ones.

- **During early spring**, winged reproductives can be seen flying and searching for an appropriate place to raise a new colony.

Termite facts

🪶 **Termites feed** on decaying wood, tree stumps, the roots of shrubs, and other vegetation. They damage anything made of wood, including furniture, doors, and windows.

🪶 **If a termite colony** becomes too big, swarming (moving in a large group) takes place. This is initiated by the winged adults of the colony, which are called alates. Swarming can also take place because of climatic changes.

🪶 **Some species** may resort to cannibalism as a means of controlling their population.

🪶 **Before swarming**, the worker termites prepare tunnels and make holes inside the nest so that the alates can fly out.

🪶 **When the alates** leave the nest, the soldiers guard the entrance and defend them from predators. They also prevent the alates from reentering.

🪶 **Alates are weak fliers** and are usually carried by the wind to different nesting grounds. This flight is called the nuptial flight.

🪶 **Once the alates** settle down in a nesting place, they shed their wings. The females produce a chemical to attract the males. The males and females then dig a hole and seal it. Mating takes place inside the hole.

▶ Giant anteaters open termite mounds using their clawed forelimbs. They catch the termites using their long, sticky tongues, which can reach 2 feet in length.

In tropical regions, the queen termites can lay eggs throughout the year. However, in temperate regions, the queens do not lay eggs in winter.

Soldier termites are blind. They defend themselves with their jaws, which function like scissors and can slice their enemies into pieces.

Termites can locate predators, such as anteaters, with their senses of touch and smell.

Termite nests

▶ Termite nests have hard walls made of soil and termite saliva, and can be domelike or conical in shape.

Termites can nest anywhere they find humid and favorable conditions. They can nest underground or in hollow tree stumps, shrub roots, cellar walls, window frames, wooden posts, and even in books.

These nests provide protection from extreme weather conditions and enemies. A large termite nest is called a termitarium.

Termites collect dirt, clay particles, and chewed wood, which they glue together with saliva or feces. They use this to build their nests.

Worker termites build the nests, which are damp, dark, and sealed from the outside environment.

The nest is ventilated by pores in the nest walls and sometimes contains tall chimneys through which warm air from inside the nest can escape.

Termite nests have many chambers, called pockets, and a network of galleries that connects the horizontal layers of chambers.

Nests can be completely underground. However, some nests rise above the ground and are known as termite hills or mounds. These hills are a regular sight in tropical regions. Certain African termites can build mounds that are 25–30 feet high.

Termites build striking nests that may be very large or have a complex architecture. They can also build cryptic nests that are difficult to find.

Ground termite

Social insects, ground termites live in large groups or colonies. They are also known as subterranean termites because they nest in soil.

Ground termites live underground. A colony can contain over a million termites, although it may be smaller in size. The maximum capacity of a colony is reached in four to five years.

In a ground termite colony, there are reproductives (king and queen), soldiers, and workers.

Once the colony reaches its full capacity, the new reproductives fly away to start their own colonies. They pair up and mate to start a new colony.

Soldiers have sharp, elongated jaws, which are used for defense against enemies, such as ants. They can tear, slice, and kill their enemies with their powerful jaws.

Workers feed the soldiers, as they cannot feed themselves. They also take care of the nest and the young ones.

Ground termites dig mud tunnels, called mud tubes, through which they travel underground to look for food. They leave behind a scent trail for other members, which contains information about the location of the food. Soldiers and workers bang their heads against tunnel walls to tell other termites about food sources and enemy attacks.

These termites have a thin outer covering called an exoskeleton. If this dries up, they die. Therefore, they always maintain contact with the moist soil around their nests.

Ground termites feed on wood and products made from wood, including books and paper. They are a serious threat to buildings and other wooden structures. People use techniques such as termite scanning and chemical treatment to keep termites away from wooden structures.

▼ *Worker termites are normally small, soft, and pale. Soldier termites defend the colony and are equipped with large heads and strong jaws.*

201

Wood termite

Wood termites are known to be very destructive, as they destroy large wooden structures from the inside. They can be a threat to houses and furniture.

These insects are also called primitive termites. There are various types of wood termites: dry-wood, damp-wood, and rotten-wood termites. They are differentiated by their nesting habits.

> **DID YOU KNOW?**
> Wood termites do not need external moisture or contact with the soil for survival. They can survive on the little moisture they get from the wood on which they nest.

Dry-wood termites can withstand the excessively dry condition of wood. They infest the dry, seasoned wood in houses and other places.

These termites eject pellets that look like wood dust from their excavations in the timber in buildings. So they are also called powder-post termites.

Damp-wood termites need constant high moisture such as is found in dead or decaying logs and the branches of live trees.

Wood termites build small colonies ranging from a few hundred to a thousand termites. Damp-wood termites usually nest in the wood, and as soon as the wood is consumed, the entire colony dies.

Rotten-wood termites produce sandlike particles that are found beside wooden structures and are a clue to their presence.

Wood termites have a unique caste system. Instead of true workers, they have false workers, which are old nymphs of the colony. They take care of the nests temporarily.

People use a chemical called termiticide to coat and protect wood from termite attacks. Fumigation is also used as a preventive measure to keep these pests away.

▼ *Wood is difficult to digest. Wood termites have bacteria and other single-celled organisms in their guts to help them digest tough plant material and wood.*

Cockroaches

Cockroaches belong to the order Blattodea, of the class Insecta. There are about 4,000 known species of cockroaches. Some well-known cockroaches are the oriental black beetles and Croton bugs.

Cockroaches are found everywhere, especially in bat caves, peoples' homes, under stones, in thick grass, and on trees and plants. The cockroaches found in caves are usually blind.

▲ Female cockroaches lay their eggs in purselike capsules. Each egg case contains about 15 eggs.

These insects may be winged or wingless. Adult cockroaches can be up to 3.5 inches long. They are nocturnal and prefer dark and damp places.

Cockroaches are swift creatures and can run extremely fast. Their legs are adapted for quick movement. They have flat, oval bodies that help them to hide in narrow cracks in walls and floors.

Most cockroach species are omnivorous. Their main food is plant sap, dead animals, and vegetable matter but they will even eat shoe polish, glue, soap, and ink.

Adults can live for up to two years. Males and females are very similar in appearance, but males have a pair of bristlelike styli.

In some species, females release pheromones or produce a hissing sound and wave their abdomen to attract male cockroaches.

Females can lay up to 30–40 eggs at a time and can reproduce four times in a year. They store the eggs in a brownish egg case called an ootheca. The cockroach can either carry this egg case along with her or hide it somewhere. Young cockroaches are called nymphs.

Cockroaches play an important role in balancing the environment by digesting a wide range of waste substances. They decompose forest and animal waste matter. However, household cockroaches can contaminate food and spread diseases among humans.

▶ Cockroaches are not dirty creatures. They work hard to keep themselves clean in order to preserve a coating of wax and oils that prevents them from drying out. It is the bacteria they carry that makes them dangerous.

Madagascan hissing cockroach

Madagascan hissing cockroaches are found on the island of Madagascar, off the southeast coast of Africa. They are large, wingless insects, famous for the loud hissing sound that they make. These hisses are loud enough for people to hear.

These cockroaches are chocolate-brown in color, with dark orange marks on their abdomen. Adults can grow up to 4 inches in length.

▼ Hissing cockroaches make their distinctive noise by pushing air out of a pair of breathing holes in the side of their body. They have spiky legs for grip and protection.

A male cockroach can distinguish between familiar males and strangers from the hissing sound that they make.

Males look different from the females. They have a large, hornlike structure behind their heads and are more aggressive than the females. They also have hairy antennae, unlike the smooth antennae of the females.

DID YOU KNOW?

Male hissing cockroaches are territorial. They fight among themselves by ramming and pushing with their horns and abdomens. Once the fight is over, the winner makes loud hissing sounds to declare victory.

These cockroaches feed on dead animal matter, waste food, or ripe fruits.

Females can produce 30–60 eggs, which they store in an egg sac, either inside or outside their body. The eggs hatch into nymphs.

The nymphs of hissing cockroaches molt six times in a period of seven months before maturing into adults. While molting, the skin of a cockroach splits down the middle of its back and the cockroach slowly wriggles out.

Newly molted cockroaches are whitish, but their color darkens within a few hours. Hissing cockroaches live for two to five years.

Madagascan hissing cockroaches can be easily bred in homes and classrooms. For this reason, they are ideal for classroom study.

Oriental cockroach

🐝 **Oriental cockroaches** are seasonal insects and can be most easily spotted during the spring and summer. They are found in damp basements, drains, leaky pipes, and kitchen sinks.

🐝 **These cockroaches** are dark brown or black in color, so they are sometimes called "black beetles." Adults can grow up to 1.5 inches in length.

🐝 **This cockroach** is sometimes called the "shad roach" because its young appear in large numbers when shad (a type of fish) are swimming into freshwater to breed.

🐝 **Males** have short wings but females are wingless.

🐝 **Females** do have small wing stubs. This is what makes them different from their own nymphs.

🐝 **A female** can produce five to ten egg cases (oothecae) in her lifetime. Each egg case contains about ten eggs.

🐝 **Young cockroach nymphs** hatch from the egg cases in six to eight weeks and mature in six to twelve months. Adult cockroaches can live for up to one year.

🐝 **Oriental cockroaches** have cerci at the end of the abdomen. These are important sensory organs.

DID YOU KNOW?

Oriental cockroaches can survive for up to one month without food, as long as water is available. Adults have a life span of about one year.

🦟 **These cockroaches** can enter homes through sewer pipes, air ducts, or any other opening.

🦟 **Unlike other pest cockroaches,** oriental cockroaches do not have sticky pads on their feet and cannot climb slippery or smooth surfaces.

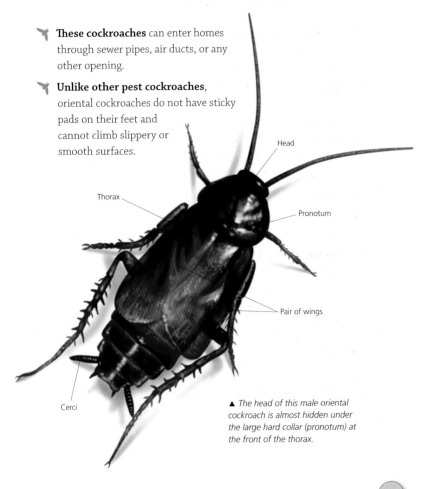

Head

Thorax

Pronotum

Pair of wings

Cerci

▲ The head of this male oriental cockroach is almost hidden under the large hard collar (pronotum) at the front of the thorax.

Death's head cockroach

▲ Adult death's head cockroaches have yellow and black markings.

- **Large insects**, death's head cockroaches measure 2-3 inches in length. These cockroaches are mostly found in the Americas.

DID YOU KNOW?
Death's head cockroaches make good pets and are popular among cockroach lovers.

- **Death's head cockroaches** are named after the strange markings on their thorax. These markings look like skulls.

- **This insect** is also known as the palmetto bug or giant death's head cockroach.

- **Death's head cockroaches live mostly** in tropical forests or bat caves, but they are sometimes found in buildings.

- **These cockroaches** are brownish in color, with yellow and black marks on their body. While adults are beautifully colored, the nymphs are dark in color, although newly molted nymphs are whitish.

- **The wings** are very long and cover their abdomen. The nymphs are wingless.

- **Females carry the eggs** inside their bodies and give birth to live young that hatch from the eggs.

- **Cockroaches are scavengers** and are active at night. Decaying plant and animal matter are their favorite foods.

Crickets

Crickets belong to the order Orthoptera and closely resemble grasshoppers. These insects are found almost everywhere—in fields, forests, and gardens.

Crickets have flattened bodies and long antennae, and measure up to 2 inches. Although all crickets have wings, some species do not fly. They can only hop from one place to another.

Male crickets make chirping sounds that attract females and warn rival males to keep away.

Crickets produce these sounds by rubbing the bases of their specially modified forewings.

▼ Crickets hatch out into nymphs that look like miniature adults without wings. They go through up to ten or more molts as they develop into winged adults.

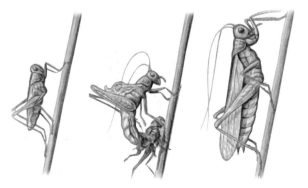

212

- **Crickets are nocturnal** and have keen hearing and eyesight. Compound eyes help crickets see far and in many directions at the same time. Round, flat hearing organs are found on the front legs.

- **Crickets are omnivorous** and feed on crops, vegetables, flowers, green plants, small animals, clothes, and each other.

- **Females** have long, needlelike ovipositors to lay eggs. They carry the eggs until they find a safe place where the eggs can hatch into nymphs.

- **Some species** of crickets are considered pests because they eat crops and deposit their eggs on them.

- **In some parts of the world**, crickets are thought to be a sign of good luck. For this reason, some people keep crickets as pets.

▲ *Crickets are common insects that are found nearly everywhere.*

Cave cricket

- **Cave crickets are named** after their habit of living in caves and other dark and damp places.

- **Cave crickets** are also known as camel crickets because they have a hump on their back.

- **Very long hind legs** give cave crickets a spiderlike appearance. Their strong hind legs make them good jumpers.

- **Cave crickets are brown** in color and have long antennae. They do not have wings.

- **This type of cricket cannot chirp** or make sounds because they are wingless. Most crickets can produce a high-pitched sound by rubbing their wings together.

- **Cave crickets are omnivorous** and feed on decaying organic matter, plants, and vegetables.

▼ The long, sensitive antennae of the cave cricket help it to find its way around in the dark. The long cerci at the end of the abdomen also have a sensory function.

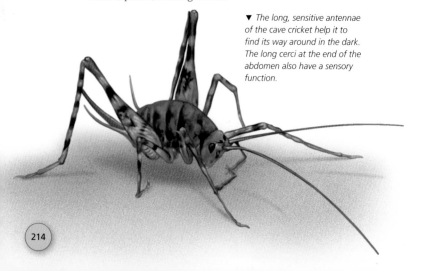

▲ *In addition to caves, cave crickets are also found in wells, rotten logs, hollow trees, and under damp leaves and stones.*

🐛 **In the spring**, females lay their eggs in soil. The nymphs and adults spend the winter in sheltered areas.

🐛 **Cave crickets** are sometimes troublesome in buildings and homes, especially in basements.

🐛 **These insects** can damage items stored in boxes, garages, and laundry rooms.

Grasshoppers

Grasshoppers are green or brown in color. Some species change colors in different seasons. The body is divided into a head, thorax, and abdomen. Grasshoppers have six legs, a pair of wings, and compound eyes.

Found almost everywhere except in the polar regions, grasshoppers prefer to live in green fields, meadows, and forest areas.

There are three types of grasshoppers—long-horned, short-horned, and pygmy. Bush crickets are also known as long-horned grasshoppers.

Grasshoppers make a loud noise during the mating season to attract a mate or scare away rivals.

▶ When a grasshopper jumps, it uses its powerful leg muscles to propel its body forward.

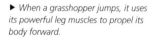

DID YOU KNOW?

When captured, grasshoppers spit a brown liquid to protect themselves from their predators. In some parts of the world, this brown liquid is known as "tobacco juice."

After mating, the female lays her eggs in low bushes or digs a hole in the soil with her abdomen to deposit eggs from her ovipositor. She covers her eggs with a hard shell covering called an eggpod.

Grasshoppers are herbivorous. They feed on a variety of plants, grasses, and crops. They use their mandibles to chew food.

These insects pose a serious threat to crops. A large group of grasshoppers can destroy an entire crop.

Flies, spiders, toads, and reptiles prey on grasshoppers and even eat their eggs.

In some parts of the world, grasshoppers are considered a delicacy. They are ground into a meal, and sometimes fried or roasted and dipped in honey.

Locust

Locusts are found all over the world except in cold regions. Dark brown in color, their average length is less than an inch.

Locusts live in fields, open woods, or arid areas. They can fly for up to 20 hours, using reserves of fat stored in their bodies.

These insects are migratory in nature and travel great distances. They can be highly destructive of crops. A large swarm can consume 3,000 tons of green plants in a single day.

Locusts breed very quickly compared to other insects. Some well-known species are the desert locust, the red-legged locust, and the Carolina locust.

Females lay their eggs in the soil. The nymphs are wingless and small in size. A female locust can lay 20 eggs at a time.

A locust eats an amount of food equivalent to its own weight every day. It feeds on crops, weeds, grass, or other plants.

◀ *A locust can use its powerful leg muscles to jump ten times its own body length.*

▲ *This photograph, taken in Morocco in 1954, shows people using poison bait in an attempt to combat locust swarms.*

DID YOU KNOW?

A locust swarm may occupy 10–300 square miles. Such swarms cast a shadow on the ground and appear like a huge black cloud.

Locusts have two phases—solitary and gregarious. In the solitary phase, locusts do not breed in large numbers and confine themselves to their surroundings. In the gregarious phase, the insects reproduce and grow quickly in the presence of an abundant food supply.

In the gregarious phase, the locusts accumulate in large numbers. As a result, the habitat where they stay is not sufficient to support and sustain them. Therefore, these insects migrate in search of new feeding grounds. This phenomenon is called swarming.

In some tribal groups, dried locusts are eaten as food.

219

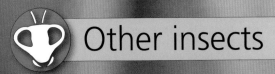

Other insects

Stick insect

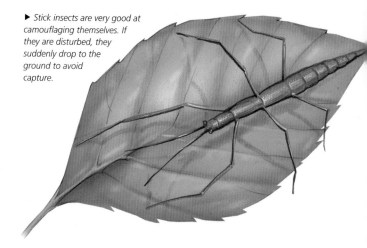

▶ Stick insects are very good at camouflaging themselves. If they are disturbed, they suddenly drop to the ground to avoid capture.

Stick insects belong to the order Phasmatodea. They look like leafless sticks and branches and are also called walking sticks.

These insects are green or brown in color and spend their time clinging to trees, plants, and shrubs.

Stick insects are excellent at camouflaging themselves, especially when they keep perfectly still or sway with the wind.

Some tropical species of stick insects have sharp spines on their legs, which blend in with the thorns of plants. If threatened, they can stab their enemies. They also change their color like chameleons, depending on the humidity and temperature.

- **Stick insects are active at night**, when they feed on foliage.

- **Females are larger** than the males. Males can fly, but the females can only glide.

- **Females produce pheromones** to attract males. A single female can lay about 1,000 eggs in her lifetime. They scatter their eggs randomly or hide them in order to protect them.

- **Asiatic stick insects** are the longest stick insects, measuring more than 12 inches.

- **Stick insects** can regenerate lost legs. If old limbs are cut off, they can grow new ones.

▶ Including its antennae and legs, the giant stick insect can measure over 20 inches in length.

DID YOU KNOW?

Stick insects lie motionless and pretend to be dead to save themselves from their predators. This is called catalepsis.

Leaf insect

- **Leaf insects belong** to the order Phasmatodea and are most common in Southeast Asia.

- **These insects** have a unique appearance. Their bodies are flat, irregular in shape, and resemble a leaf. These insects are about 4 inches in length and sometimes have brown or yellow patches on their body.

- **Females** are much larger than males. They do not have hind wings and cannot fly, but they do have leaflike forewings.

- **The female can lay eggs** without mating with the male. This is called parthenogenesis.

- **A female** scatters her eggs on the ground. These eggs have a hard outer shell and resemble seeds, so predators do not feed on them.

- **The eggs** hatch in the spring.

- **Leaf insects are herbivorous** and feed only on green leaves and other parts of plants.

- **Rodents**, birds, and other insects feed on leaf insects.

- **Leaf insects** are good at camouflage and hiding from predators. They move very slowly and mimic the foliage on which they live.

▶ *This leaf insect would be very hard for predators to spot among green leaves, as long as it remained completely still.*

Damselfly

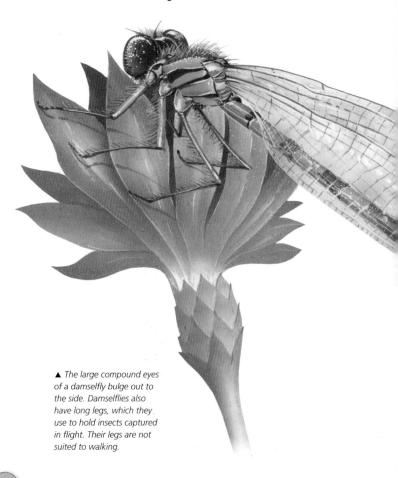

▲ The large compound eyes
of a damselfly bulge out to
the side. Damselflies also
have long legs, which they
use to hold insects captured
in flight. Their legs are not
suited to walking.

- **Damselflies** belong to the order Odonata and suborder Zygoptera. They have long, slender bodies and four long wings. They are weak fliers and timid predators.

- **These insects are beautiful**, slender cousins of dragonflies but are not as ferocious. They have compound eyes and excellent eyesight.

DID YOU KNOW?
Damselflies capture their prey with the help of their legs, folding them like a basket to form a trap. Once the insect is trapped it is transferred to the damselfly's mouth.

- **Damselflies** are usually found near water, and their nymphs live in the water until they mature into adults.

- **Mosquitoes**, midges, gnats, and small water insects are the preferred food of damselflies.

- **Damselfly nymphs** have external gills on the tip of their abdomen for breathing underwater. In dragonfly nymphs, these gills are internal.

- **Males and females** mate during flight or over shallow water. While mating, a male carries a female around to allow her to collect sperm from the front of his abdomen. After mating, females deposit their eggs in and around water.

- **The life span** of a damselfly is around one year, but it can live for up to two years.

- **During winter**, damselflies hibernate to survive the cold weather.

227

Dragonfly

- **Dragonflies are named** for their fierce jaws, which they use for catching prey.

- **These insects can grow up** to 5 inches in length. They have long, slender, multicolored bodies with two pairs of veined wings.

- **Larger dragonflies** are called hawkers, while the smaller ones are called darters.

- **Adult dragonflies** survive on land, but their nymphs live underwater. They cannot survive harsh winters.

▶ Dragonflies have huge compound eyes, which cover their entire head.

DID YOU KNOW?

Dragonflies are one of the fastest insects. They can fly up to 30 mph. The design and mechanism of a helicopter was probably inspired by a dragonfly.

The life span of a dragonfly ranges from six months to more than seven years.

Males and females mate while they fly. Once they have mated, the female dragonfly deposits her eggs in water or inside water plants.

A female dragonfly lays up to 100,000 eggs at a time. The eggs hatch into nymphs, which feed on fish, tadpoles, and other small aquatic animals.

Dragonflies are beneficial to humans. They prey on mosquitoes, flies, and many small insects that are pests.

Experts have found dragonfly fossils that are more than 300 million years old. These prehistoric dragonflies were as big as crows.

Lacewing

🦋 **Lacewings** are delicate green or brown insects that are named after their thin, translucent, lacelike wings.

🦋 **These insects have long**, threadlike antennae and their eyes are bright golden yellow or brown in color.

🦋 **Birds and other predators** avoid lacewings because they give off a pungent odor similar to garlic.

🦋 **Lacewings are not good fliers** because their wings are too weak to support their weight for long distances. These insects are often seen fluttering about clumsily.

🦋 **A lacewing lays** a large number of eggs at a time. The eggs are white in color and are glued to a twig or leaf. Green lacewings secrete a thin white stalk for each egg.

🦋 **Lacewing larvae** feed on small insects. They eat a lot of aphids and are also known as aphid lions. These insects are sometimes used to control the population of aphids in crop fields.

🦋 **The larvae of lacewings** are sometimes reared commercially to get rid of pests. Apart from aphids, these larvae feed on mealybugs, leafhoppers, caterpillars, moths, and other insects. The larvae can become cannibalistic if there is not enough food.

🦋 **Lacewing larvae** have "jaws," which make them resemble alligators. These jaws are used to suck the body fluids of the prey after injecting it with paralyzing venom.

▲ *A lacewing has two similar pairs of wings, which are covered with a delicate network of veins. When the lacewing rests, it holds its wings together like a roof over its body.*

Some lacewing larvae camouflage themselves with debris, including the skin and other remains of their prey.

Many species of adult lacewings do not eat other insects.

Praying mantis

Closely related to cockroaches, praying mantises belong to the order Mantodea.

Praying mantises get their name from their posture. This insect holds its front legs together as if it were praying.

▼ *A praying mantis moves its long front legs very rapidly to grab its prey. Sharp spines stop the prey from escaping.*

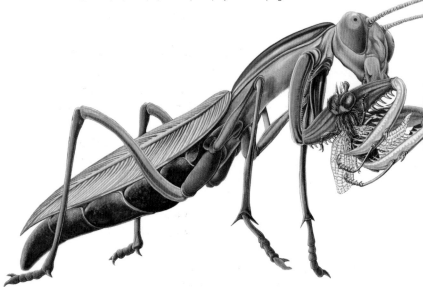

DID YOU KNOW?
A female praying mantis often eats the male while mating!

🦟 **These insects** can grow up to 2.5 inches in length. They have triangular heads that can turn in a full circle. They are generally green or brown.

🦟 **Praying mantises have good** camouflage to protect themselves from predators. Their body color blends with their environment. A species called the Asiatic rose mantis even looks like flower petals. For this reason, it is also called the orchid mantis.

🦟 **Females are larger** than the males. They lay their eggs inside an egg case (ootheca) during the autumn.

🦟 **Females secrete** a sticky substance to stick their eggs to plant stems and tree twigs. Nymphs hatch from these eggs in the spring or summer.

🦟 **Praying mantises** are carnivorous and feed on insects such as butterflies, moths, and grasshoppers. They can even attack small lizards, frogs, and birds. Small mantises can become cannibalistic, especially when there is no food.

🦟 **These insects** have a habit of silently watching and stalking prey before attacking. While feeding, a praying mantis holds its prey with its front legs.

🦟 **Praying mantises are useful** to humans. They protect crops by feeding on pests. Sometimes farmers buy them for pest control. These insects are often kept as pets, even though large praying mantises have a painful bite.

Mantisfly

Mantisflies are named after their folded pair of front legs, which look like those of praying mantises. Although these insects look similar, they are not related.

These insects are not common and are found in tropical regions.

Mantisflies, along with lacewings and ant lions, belong to the order Neuroptera.

Mantisflies lay their rose-colored eggs on slender plant stalks.

The mantisfly has a complicated development cycle. It has two growth stages as a larva and two more stages as a pupa.

One species of mantisfly has larvae that are a parasite of wolf spiders.

A single larva enters the spider's egg sac and preys upon the egg or young.

The larva pierces the egg or spider with its pointed mouthparts and feeds on its body fluids. The larva then pupates inside the spider's egg sac.

Unaware of the presence of the parasite mantisfly, the parent spider watches over the egg sac.

Some mantisflies mimic wasps and bees for protection.

Mantisflies are predatory like praying mantises and have similar feeding habits.

◄ A mantisfly grabs its prey with its huge front legs. The wings of a mantisfly are very different from those of a praying mantis.

Ant lion

- **Ant lions are named** for the larva's habit of feeding on ants and other insects. Adult ant lions feed on pollen or nectar, or do not feed at all.

- **Ant lions resemble dragonflies**. They both have long, slender bodies and four delicate wings. However, unlike dragonflies, ant lions are nocturnal insects.

- **These insects** are found in damp areas where vegetation is thick. Ant lions are also found near riverbeds.

- **Some ant lion larvae** make a pit in the sand by moving around in a spiral path and throwing out sand.

- **While making the pit**, the larva leaves behind squiggly doodles on the sand. For this reason, ant lions are also known as doodlebugs.

▼ An antlion larva waits at the bottom of its pit trap for an unsuspecting ant to fall down into its pincerlike jaws.

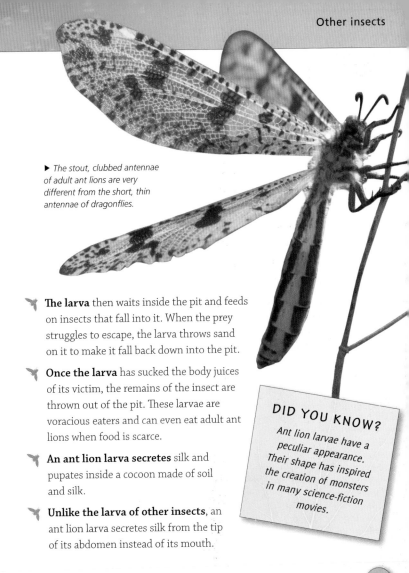

▶ The stout, clubbed antennae of adult ant lions are very different from the short, thin antennae of dragonflies.

The larva then waits inside the pit and feeds on insects that fall into it. When the prey struggles to escape, the larva throws sand on it to make it fall back down into the pit.

Once the larva has sucked the body juices of its victim, the remains of the insect are thrown out of the pit. These larvae are voracious eaters and can even eat adult ant lions when food is scarce.

An ant lion larva secretes silk and pupates inside a cocoon made of soil and silk.

Unlike the larva of other insects, an ant lion larva secretes silk from the tip of its abdomen instead of its mouth.

DID YOU KNOW?
Ant lion larvae have a peculiar appearance. Their shape has inspired the creation of monsters in many science-fiction movies.

237

Stone fly

🐝 **Stone flies are an ancient order** of insects that has been around for nearly 300 million years.

🐝 **Stone fly nymphs** cling to stones in clear mountain streams or lakes. They are eaten by fish and are also used as fish bait.

🐝 **Some stone fly nymphs** breathe through gills that are present near their legs. Others obtain oxygen through their body surface.

🐝 **Adult stone flies** are terrestrial, which means that they live on land. However, they never wander too far from water, spending their time crawling over stones. This is how they got their name.

🐝 **Female stone flies** deposit their eggs in deep water. To prevent the eggs from floating away, they stick them to rocks with a sticky secretion.

> DID YOU KNOW?
> Some male stone flies attract females by drumming their abdomen against a hard surface. For this reason, these insects are popularly known as "primitive drummers."

▼ *A stone fly lives for up to two or three years.*

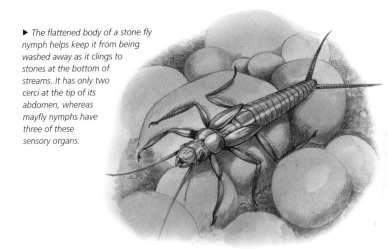

▶ The flattened body of a stone fly nymph helps keep it from being washed away as it clings to stones at the bottom of streams. It has only two cerci at the tip of its abdomen, whereas mayfly nymphs have three of these sensory organs.

A stone fly nymph resembles an adult but does not have wings, and its reproductive organs are not as well developed. These nymphs grow into adult stone flies and crawl out of the water.

Stone fly nymphs feed on underwater plants such as algae and lichens. Some species are carnivorous and feed on aquatic insects. Adults may not feed at all, although some eat algae and pollen.

Fish often attack female stone flies when they try to deposit their eggs in streams.

Stone flies are sensitive to water pollution. Experts use these insects to study the level of water purity.

Scorpion fly

Scorpion flies are a small group of insects with dark spots on their two pairs of membranous wings. These primitive insects have been around for 250 million years.

The rear body part of some males is curled upward, similar to a scorpion's. Only males resemble scorpions.

Adult scorpion flies have modified mouthparts. They have a long beak that points downward.

Scorpion flies feed on live as well as dead insects. Some also feed on pollen, nectar, and plants.

DID YOU KNOW?
During the mating season, a male scorpion fly offers the female a gift. This courtship gift can be a dead insect or a drop of his saliva.

▼ Male scorpion flies have a large stinger on a long, curved tail. It looks like the stinger of a scorpion but is harmless.

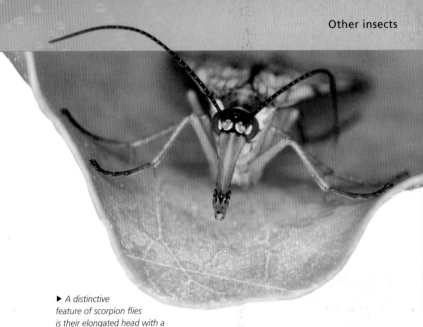

▶ A distinctive feature of scorpion flies is their elongated head with a "beak" that ends in biting jaws.

🪰 **Scorpion flies** are predators and often steal their food from spiderwebs. They catch their prey with their hind legs.

🪰 **While mating**, a male scorpion fly secretes a sweet, sticky fluid, on which the female scorpion fly feeds.

🪰 **The female** lays her eggs in wooden crevices and soil. The eggs hatch and the larvae live and pupate in loose soil or waste matter.

🪰 **Scorpion fly larvae feed** on dead and rotten plant and animal matter. Some larvae also feed on small insects.

🪰 **Scorpion flies** are often considered helpful to humans because they keep the environment clean by feeding on dead insects.

Mayfly

▲ *Adult mayflies have pale yellow bodies with brown stripes.*

Mayflies belong to the order Ephemeroptera—*ephemeros* means "one day" and *pteron* means "wing." They are slender insects and measure up to 1.5 inches in length.

Easily spotted around streams and ponds, mayflies are characterized by triangular, membranous forewings and smaller, rounder hind wings.

Mayflies have soft, fragile bodies and also have two or three long, threadlike tails.

- **Even after** their wings become functional, mayflies molt. This is the preadult stage.

- **While resting**, mayflies hold their wings vertically and not in a rooflike position as other insects do.

- **Male mayflies** "dance" in large swarms to attract females.

- **A female mayfly** lays her eggs in water by dipping her abdomen into the water while in flight.

- **Strongly attracted** to light, mayflies gather around lights on roads and streets.

- **Adult mayflies** do not bite humans, but their huge swarms are a nuisance.

DID YOU KNOW?
The life span of an adult mayfly is short and ranges from a few hours to a few days. It lives just long enough to mate and reproduce, and does not feed.

▼ Mayfly nymphs are good swimmers and live for two to three years underwater. They mostly feed on plant matter that has settled at the bottom of the river.

243

Flea

🪶 **Fleas belong** to the order Siphonaptera. They are ectoparasites—parasites that live on the outside of their host.

🪶 **Small, wingless insects**, fleas measure 0.1–1 cm in length. These insects feed only on the blood of birds and mammals, including humans.

🪶 **Bloodsucking pests**, fleas carry various deadly diseases. They transmit bubonic plague, which killed half the population of Europe in the mid-1300s.

🪶 **When rats** die of bubonic plague, fleas infected with the plague bacilli seek food elsewhere. These fleas may transmit the disease to humans.

🪶 **Fleas do not infest** monkeys, horses, or apes.

🪶 **Infestation by fleas** causes severe inflammation of the skin and itching. Animals heavily infested by fleas may die of a fleabite or severe blood loss.

🪶 **Fleas** are themselves subject to parasitism by external mites and internal nematode worms.

DID YOU KNOW?
Flea larvae may remain in their cocoons for months until they sense the movement of an animal nearby.

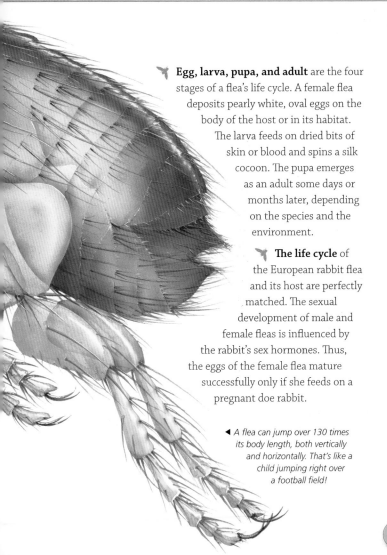

Egg, larva, pupa, and adult are the four stages of a flea's life cycle. A female flea deposits pearly white, oval eggs on the body of the host or in its habitat. The larva feeds on dried bits of skin or blood and spins a silk cocoon. The pupa emerges as an adult some days or months later, depending on the species and the environment.

The life cycle of the European rabbit flea and its host are perfectly matched. The sexual development of male and female fleas is influenced by the rabbit's sex hormones. Thus, the eggs of the female flea mature successfully only if she feeds on a pregnant doe rabbit.

◀ A flea can jump over 130 times its body length, both vertically and horizontally. That's like a child jumping right over a football field!

245

Lice

🦟 **Many parasites** are capable of living off humans. They can infest hair, body, or blood. Lice are just one type of parasite that thrives on the human body.

🦟 **Human lice belong** to the order Anoplura and a group known as sucking lice.

🦟 **Lice are small, wingless insects** measuring up to half an inch long. Two types of lice that frequently affect humans are the head louse and the body louse.

🦟 **The head louse** commonly infests children but is also found on adults. The easiest way to get rid of them is to comb the hair thoroughly with a very fine "nit" comb and use a treatment such as tea tree oil. Lice are more common on clean hair.

🦟 **A louse's body** is flattened, which helps it to lie close to the skin. The head louse attaches itself to the hair or scalp with the help of claws on its legs.

🦟 **Lice feed** on human blood. They have three needlelike structures, or stylets, which can pierce the skin. These are held back by a specially adapted tongue.

🦟 **After piercing** the skin, the lice suck the blood with a pumping action of their throat. When the stylets are not in use, they are retracted into a pocket just behind the louse's mouth.

🦟 **Lice frequently feed on blood,** and each feeding takes a few minutes.

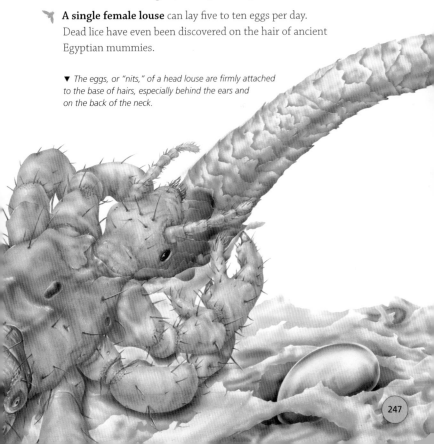

An adult head louse may also be found in any facial hair, but they are almost never found below the neck. Head lice are transmitted by direct contact with an infested person or by using infested articles such as headgear, combs, brushes, or scarves.

A single female louse can lay five to ten eggs per day. Dead lice have even been discovered on the hair of ancient Egyptian mummies.

▼ *The eggs, or "nits," of a head louse are firmly attached to the base of hairs, especially behind the ears and on the back of the neck.*

247

Book louse

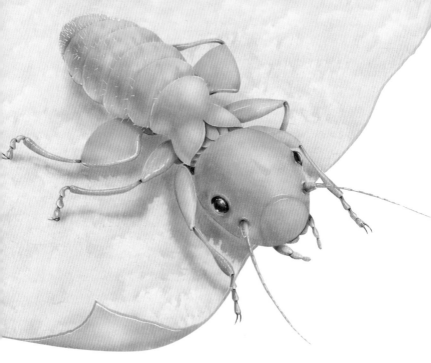

▲ Book lice are so small that they are almost invisible to the naked eye. They can, however, destroy valuable collections of books by eating the glue that binds the pages.

Book lice are small, colorless insects measuring 1–4 mm. They look like lice but are not true lice.

- **Soft-bodied**, transparent, and usually wingless, they infest books, paper, and cardboard boxes, especially those that are stored in dark and damp places. Book lice belong to the order Psocoptera.

- **The head and abdomen** appear large, while the thorax is narrow. Book lice have long, threadlike antennae and chewing mouthparts. Their eyes are large and protrude from the sides of the head.

- **Book lice** do not bite humans or animals. Although they do not spread diseases or damage household furnishings, they often damage books and papers.

- **Also known as paper or dust lice**, they can be spotted in large numbers near stored books and papers in spring or summer.

- **These insects feed** on microscopic mold, fungi, dead insect fragments, pollen, and other starchy food scraps and mildew that are found in a humid environment.

- **Book lice** run in a jerky, halting manner and sometimes appear to hop.

- **Their life cycle**, from egg to adult, takes from four weeks to two months or more, depending on the species and environmental conditions.

- **The outdoor species** of book lice are known as bark lice, as they are found under tree bark or leaves. They feed on plant fiber. Bark lice are also found in birds' nests.

- **Bark lice live** in groups. They may be winged or wingless. The winged species, however, are weak fliers.

Earwig

Earwigs belong to the order Dermaptera, meaning "skin wings." They measure about 1 inch and are flat and slender. Their color ranges from reddish-brown to black. They have simple, biting mouthparts.

These insects usually have membranous hind wings, which lie hidden under their leathery forewings. Some species of earwigs are wingless.

The hind wings open in a shape that resembles that of a human ear.

Earwigs have a pair of horned forceps, called cerci, at the tip of their abdomen. The cerci look like tails or pincers.

Cerci help earwigs to defend themselves. When alarmed, earwigs hold their cerci over their body much like a scorpion does. Although the cerci give earwigs a fierce look and can inflict sharp bites, they are quite harmless to humans.

DID YOU KNOW?

One superstition is that earwigs crawl into the ears of people while they are asleep. This is how they got their name. Despite their threatening appearance and worrying name, earwigs are unlikely to come near a human, let alone crawl into the ears!

The male has longer and more curved pincers than the female.

Several species of earwigs release a foul-smelling liquid, which is produced in their abdominal glands. This helps to protect them from predators.

🪰 **The female** lays 25–30 eggs below the ground. She incubates them like a brooding hen. These eggs hatch into nymphs and gradually develop into adults.

🪰 **Earwigs feed** on algae, fungi, mosses, pollen, dead and live insects, spiders, and mites. They damage flowers, vegetables, and fruits by feeding on them.

▼ *Earwigs are nocturnal insects. They live in the soil and can dig tunnels as deep as 6 feet under the ground in order to escape cold weather.*

Silverfish

Silverfish are primitive insects that belong to the order Thysanura.

These insects are soft-bodied and do not have wings. Silverfish have a longish body with three tail filaments.

▼ Silverfish are named after the tiny, shiny scales that cover their carrot-shaped body. They belong to a group of insects called bristletails because their abdomen ends in a central tail, which is fringed with bristles.

- **Silver or brown** in color, silverfish can move very swiftly.

- **Silverfish prefer to live in** places that are dark and humid. They are mostly found in moist and warm places, such as kitchens and baths.

DID YOU KNOW?
Silverfish are able to squeeze their flattened bodies into the smallest of gaps.

- **These insects generally feed** on flour, glue, paper, leftover food, and even clothes. Silverfish can easily survive for months without any food.

- **Silverfish mate** in a unique manner. A male spins a silk thread and deposits his sperm on it. The female then comes near this thread, picks up the sperm, and uses it to fertilize her eggs.

- **A female silverfish** can lay up to 100 eggs in her lifetime. After mating, she deposits clusters of her eggs in cracks or crevices.

- **Even after** maturing into adults, these insects continue to molt. They can live for two to eight years.

- **Generally harmless** to humans, silverfish can contaminate food.

- **Silverfish can destroy** books and are considered indoor pests.

Springtail

- **Springtails** belong to the order Collembola.

- **Wingless insects**, springtails measure 1–5 mm in length. Unlike most other insects, springtails do not have compound eyes.

- **These insects** have a springlike organ, known as a furcula, under their abdomen. It helps them to leap high into the air.

- **Springtails** do not have a respiratory system. They breathe through their cuticle (hard skin).

- **Even though springtails can jump**, they normally crawl from one place to another.

- **Springtails prefer** to live in soil and moist habitats. However, they can survive almost anywhere in the world, including Antarctica and the Arctic.

DID YOU KNOW?

Springtails are also called snow fleas as they can survive in extreme cold. They are active even in freezing weather.

◀ A springtail's forked tail is held under tension below the body. When released, it pings down and back to fling the springtail away from danger.

🪰 **Decaying vegetable matter**, pollen, algae, and other plants are the preferred food of springtails.

🪰 **A female springtail** lays approximately 90–150 eggs in her lifetime.

🪰 **The life span** of a springtail is one year or less.

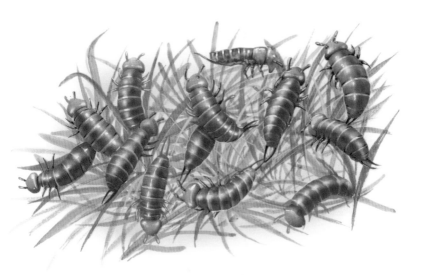

▲ *Springtails have to live in moist places because their bodies dry out easily. They are usually a gray or brown color, sometimes with mottled colors for camouflage.*

Proturans

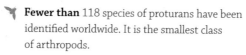

Proturans are small hexapods (with six pairs of legs) measuring 0.5–2 mm in length. However, they do not have eyes, wings, or antennae.

Proturans belong to the order Protura and are primitive insects that live in soil.

Their abdomen is divided into 12 segments, more than any other insect. They are the most primitive hexapods, and some experts consider proturans to be very different from true insects.

Proturans were first discovered in Italy but have since been found in Europe, India, the United States, and Canada.

Fewer than 118 species of proturans have been identified worldwide. It is the smallest class of arthropods.

Found in damp, dark, moist places, proturans feed on decaying matter and other insects.

Proturans' mouthparts are modified for piercing and sucking.

These insects have a unique tail, which is used for movement and defense. When threatened, they curve their tails over their heads and discharge a sticky secretion on their enemies.

Proturans use their first pair of legs as antennae.

DID YOU KNOW?

Proturan larvae have only a few abdominal segments, but with molting, new segments are added to their abdomen. This process is known as anamorphosis. This is unheard of in other insects but occurs in millipedes, centipedes, and annelid worms.

▲ Proturans have neither eyes nor antennae. Instead, they have a sensory organ on their head and use their front two legs as antennae. Most species are microscopic.

Diplurans

Diplurans belong to the order Diplura. They are small insects and are usually less than a quarter of an inch in length. However, some species can grow to 2 inches in length.

Although rarely spotted, diplurans are common arthropods and there are more than 200 species worldwide. They are related to proturans and springtails.

▼ Diplurans can be easily spotted in the wet forests of Australia. They are often found in small groups in the soil or under rocks or the bark of trees. Some species also live in ant and termite nests.

These insects are blind. Their abdominal cerci look like pincers, and they are often mistaken for earwigs.

In some species, the cerci break off near the base if they are mishandled.

A tiny packet of sperm, known as spermatophore, is deposited by the male. When a female comes across this, she fertilizes herself with it. This form of fertilization is known as external fertilization.

The female lays her eggs in crevices, in the soil, or on rotting vegetation. Some species are known to protect their eggs and young larvae.

Molting continues throughout their short life span of one year. An adult may molt up to 30 times.

Most species feed on a variety of plant matter. Diplurans that have pincerlike cerci are carnivorous. They bury themselves in the soil, with their tail on the surface. They wait for insects and capture them with their cerci.

Small vesicles (sacs) on their abdominal segments help them to maintain the water balance in their body by absorbing moisture from the environment.

As diplurans live in the soil and feed on plant tissues and decaying vegetation, they often damage plants by feeding on them.

 Spiders, scorpions, and mites

Arachnid anatomy

Arachnids include a variety of creatures, such as spiders, scorpions, ticks, mites, harvestmen, and schizomids. Fossils of arachnids suggest that they were among the first land inhabitants of this planet. Arachnids can be found anywhere, but they are most common in dry and tropical regions.

These creatures have eight legs. They also have two pairs of appendages (the chelicerae and the pedipalps) at the front of the body, which are used to grasp and hold prey.

Arachnids are classified as invertebrates—animals that do not have backbones. Instead of lungs, arachnids have two types of breathing mechanisms—book lungs and tracheae.

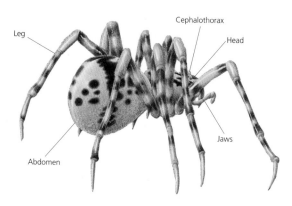

Cephalothorax

Leg

Head

Abdomen

Jaws

▲ *Spiders are one of the best-known arachnids. The abdomen of a spider contains silk glands.*

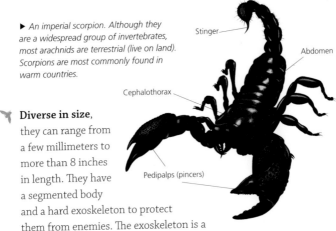

▶ *An imperial scorpion. Although they are a widespread group of invertebrates, most arachnids are terrestrial (live on land). Scorpions are most commonly found in warm countries.*

Stinger

Abdomen

Cephalothorax

Pedipalps (pincers)

Diverse in size, they can range from a few millimeters to more than 8 inches in length. They have a segmented body and a hard exoskeleton to protect them from enemies. The exoskeleton is a shell made of chitin and calcium.

An arachnid's body is divided into two parts: the cephalothorax (joint head and thorax) and the abdomen. The cephalothorax has sensory organs, mouthparts, stomach, and limbs, while the abdomen contains the heart, lungs, gut, reproductive organs, and anus.

Arachnids do not have teeth and jaws to chew their food, and most cannot digest food. This is why they suck fluids from their prey's body.

Sensory hairs, simple eyes, and slit sensory organs are used to sense the surroundings.

Arachnids are cold-blooded creatures and get warmth from their environment.

Arachnids

🕷 **Arachnids are ecologically important** to humans, as they help to control the populations of pest insects.

🕷 **The life span** of arachnids is fairly short. In temperate regions, they usually live for one year, but in warm places their life span can be longer.

🕷 **When arachnids walk,** they use the first and third legs on one side of their body and the second and fourth legs on the other side. These legs work in tandem to help them move.

🕷 **Certain species** attach themselves to the legs of other moving animals in order to move around. This mode of movement is called phoresy.

🕷 **Some mites** travel from one place to another by floating on the wind.

🕷 **Baby spiders** move from one place to another using a silken strand released from their body. The wind then carries this thread and the spiders hang on for a free ride. This technique is called ballooning.

🕷 **Ticks** are the most dangerous parasites of humans. Their main diet is the blood of mammals, reptiles, and birds.

🕷 **Some species** of ticks use their saliva to stick to living things so that they can suck blood. Once a tick is attached, it is very difficult to remove.

🕷 **When caught** by predators, some arachnids drop their captured limbs in order to escape.

Arachnids are feared by humans because of the painful stings of scorpions, the venomous bites of some spiders, and the diseases caused and spread by mites and ticks.

▼ *A microscopic tick. There are estimated to be about 75,000 species of arachnids, including scorpions, spiders, ticks, and mites.*

Spiders

Spiders, along with scorpions, ticks, and mites, belong to the family Aranaea within the class Arachnida. They differ from insects in that they have eight legs.

Found everywhere in the world except in very cold places such as Antarctica, spiders can survive in dry as well as moist areas.

Males are smaller than females. They may also have different markings or coloring.

▼ The Australian redback spider is one of the most deadly of a group called widow spiders. These spiders get their name because once they have mated, the female eats the male.

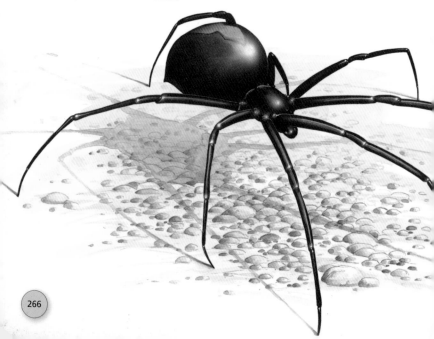

When the male finds the right partner, he usually mates with her immediately. However, some species secrete pheromones (chemical signals) to attract a mate. These spiders also have elaborate mating rituals.

Some females eat the male after mating. The male is often too weak to defend himself from this attack.

The female lays her eggs inside egg sacs made of silk threads. Each sac may contain hundreds of spider eggs. The sacs are either carried around by the mother spider or hidden in her web.

Spider eggs take a few weeks to hatch. Small spiders come out of the eggs. These are known as spiderlings. Young spiders may live in a group until they mature, but some spiderlings live alone.

Spiderlings molt a couple of times before becoming fully grown adults.

Most species are solitary and only come together to mate. Some spiders, however, live as a community in which a group of females share one web.

▶ Spiders are predatory creatures. About half of the 40,000 known spiders weave webs or nets to catch prey. They can sense the vibrations when an insect rests on the web.

267

More spiders

Experts have found spider fossils that are 380 million years old.

Spiders are carnivores. They feed on all types of insects. Some bigger species can even attack small mammals and birds.

Most spiders inject paralyzing venom into their trapped prey. The prey is partly digested by this venom, and then the spider sucks out its fluids.

Some spiders are very good hunters. They go in search of insects or wait and make a surprise attack on them.

Moth pheromones are secreted by some spiders. These chemicals attract male moths, which are fooled into thinking that the spider is a female moth waiting to mate. The spider then kills and eats the moth.

Many spiders are well camouflaged by their colors and patterns. This, along with their jumping skills, helps spiders to protect themselves from enemies.

Spider wasps are one of the main predators of spiders. However, scorpions, centipedes, small mammals, and birds are also enemies.

DID YOU KNOW?

Out of the 40,000 different kinds of spiders, only about 30 are dangerous to people.

Spiders are important to the environment. They are good pest controllers. Useful to humans, spiders prey on the pests that attack rice fields and cotton, apple, and banana plantations.

▶ Many spiders have small bodies in comparison to their long, very powerful legs.

Spider stories

The scientific name for spiders as a group is arachnids. This term is derived from an ancient legend about a beautiful Greek lady called Arachne.

Arachne was very talented and could weave beautiful patterns on cloth. Her work was so good that she challenged the Greek goddess Athene to compete with her.

Athene accepted the challenge, but Arachne wove a much better pattern than Athene and won the competition. The goddess was very angry and tore Arachne's work into pieces. Arachne was so sad that she tried to hang herself. The goddess then felt sorry for Arachne and changed her into a spider so that she could continue weaving her beautiful patterns in the form of webs.

◀ Some of the best known spider stories come from West Africa and the Caribbean. They describe the adventures of Anansi, a clever, cunning spider – and are an encouragement to children to be clever.

Native Americans believed that Spider Woman was the creator of the universe as we see it today. She made the skies, the sun, and the moon. Spider Woman was also seen as the mother of all wisdom.

During the medieval period, Europeans believed that spiders caused madness and mass hysteria. They thought that people who were bitten by spiders would suffer from pain and laugh or cry for no reason.

Medieval Europeans also believed that music and dancing was the only way to cure the affected people. They later found out that spider bites had nothing to do with madness.

King Robert the Bruce of Scotland fought many wars against England and lost all of them. He hid from his enemies inside a cave and felt hopeless.

Legend has it that the king was inspired by a spider to fight once more against England. This spider kept trying to weave a web and did not give up, in spite of failing many times. King Robert the Bruce went on to win the next war against England.

The Prophet Muhammad, the founder of Islam, hid inside a cave away from his enemies. A spider spun a web across the cave's entrance and prevented enemies from spotting the Prophet.

DID YOU KNOW?

West African mythology has a very famous character named Anansi the spider. There are many tales about Anansi and his adventures. In the stories, the spider is described as being very clever and bringing fire, food, and water for human beings from heaven.

Defense strategies

Spiders prefer to live in secluded areas and avoid trouble. However, these arachnids can become aggressive if attacked.

Special glands in a spider's body produce venom. They use venom to protect themselves and to capture prey. One spider family (Uloboridae) does not have venom glands.

Venom glands are located near the fangs on the spider's head. If attacked, spiders stand on their hind legs and raise their head to display the fangs.

Some animals can be killed with spider venom, while others are paralyzed. Spider venom can affect animals in two ways—it can be neurotoxic or necrotic.

Neurotoxic venom affects the entire nervous system of the animal. It can cause paralysis and pain in areas other than where the spider has bitten. Necrotic venom affects only the tissues where the spider bites. The skin in this area can form blisters or turn black.

Apart from using venom, spiders protect themselves with the help of camouflage. Some spiders resemble bird droppings, while others look like dried leaves. These spiders avoid being noticed by predators.

Other spiders, such as some crab spiders, can even change color to suit their surroundings.

▶ *A female tarantula guarding her eggs. The eggs are surrounded by a shell, or cocoon, of silk. Tarantulas may look threatening but are in fact shy creatures with weak venom that is not strong enough to kill humans.*

Spiders are hairy creatures. The hair may help with defense. When they are threatened, tarantulas hurl tufts of barbed hairs at their attacker. This leads to skin irritation and pain.

Sometimes spiders pretend to be dead if they are attacked. Predators prefer to eat live prey, so they leave the spider alone.

Most spiders do not cause harm to people. However, the venom of brown recluse spiders, black widows, and Brazilian wandering spiders can kill people.

Silk

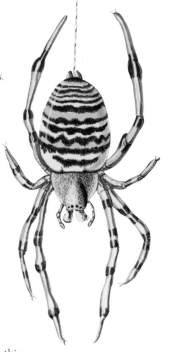

▶ White-backed garden spiders have distinctive silver, yellow, and black stripes. Like their close relatives, yellow garden spiders, they build large webs and often hang upside down near the centers.

Spiders specialize in secreting silk. They have special organs to produce silk. These organs, which are known as spinnerets, are located at the tip of the abdomen.

Spinnerets have lots of openings, called spigots, through which silk comes out. Each spigot secretes silk of a different thickness.

Spider silk is made of protein particles. Inside the spider's body, this silk is liquid in form.

The liquid silk hardens as the spider pulls it out to form a fine thread. Spider silk is very elastic and stretchy.

Many glands in a spider's body produce silk. Some spiders have seven silk glands. Different types of silk are used for different purposes. Sticky silk is used to capture prey, while nonstick silk is used to make webs strong.

🐦 **Silk is used** to wrap the prey, to travel long distances via ballooning, and to make special sacs for spider eggs. Males also use silk to form a parcel of sperm.

🐦 **Some spiders**, such as tarantulas, do not spin webs, but they do line their burrows with silk.

🐦 **Spiders** normally weave webs during the night and often finish making them by early morning.

🐦 **Spider silk** has been used in a variety of optical devices, such as microscopes. Its strength and capacity to adapt to various climatic conditions makes spider silk suitable for such devices.

> ## DID YOU KNOW?
> To avoid getting stuck inside their own webs, spiders smear a special substance over their bodies and have oil glands on their feet. They only tiptoe while moving on the web. A spider also knows where it has laid the sticky and non-sticky silk threads.

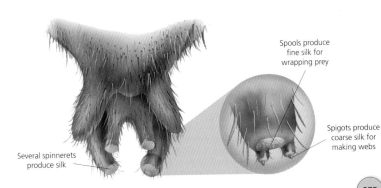

Spools produce fine silk for wrapping prey

Spigots produce coarse silk for making webs

Several spinnerets produce silk

275

Webs

Spiderwebs are of different shapes and sizes. The webs can be grouped into three main shapes—sheet, orb, and spatial. The most common are orb-shaped.

Spider silk is made from protein, and it is stronger than a piece of steel of the same thickness.

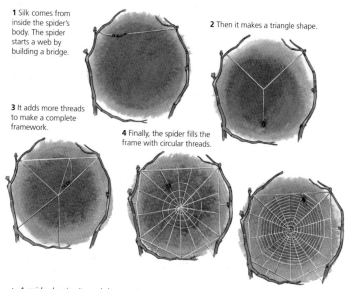

1 Silk comes from inside the spider's body. The spider starts a web by building a bridge.

2 Then it makes a triangle shape.

3 It adds more threads to make a complete framework.

4 Finally, the spider fills the frame with circular threads.

▲ A spider begins its web by creating a strong framework of silk. It then fills the frame with sticky, circular threads that trap prey such as flies.

5 A spiderweb is strong enough to catch large insects, but it is easily damaged by larger animals and people.

Spiders depend on wind currents to lay the first thread of the web. They float the thread in the air and wait for it to stick to a plant stalk or some other support. Then the spider walks across the thread to continue weaving the web.

Orb webs are very light but they are made in such a way that the delicate web can support the spider as well as the struggling prey. This means that the web can handle a weight that is thousands of times heavier than itself.

Some spiders weave a web that resembles a hammock. The web is spun across two plants and the insect flies into it. The spider waits beneath the web for the prey to be caught and fall. Any holes made on the web are repaired later.

Sheet webs are also common among spiders. The spider weaves a sheet of sticky silk and lays it flat. Right above the sheet, it weaves a number of nonstick silk threads. Insects that fly past the web will trip on the nonstick threads and fall into the sheet below.

Not all spiderwebs are built against supporting structures. Some spiders use the web like a net. The net is thrown over the unsuspecting prey, which is then wrapped in silk.

DID YOU KNOW?
Certain spiders weave elegant webs and make special "rooms" with doors, which they hide behind. They may also hide at a distance and only come into the web after an insect is trapped.

House spider

- **House spiders** are the most common spiders. They are easily spotted in houses and gardens. They are also found in woodpiles and under logs.

- **These spiders** have a yellowish-brown body with gray and black markings on their abdomen.

- **Males are smaller** than females but have longer legs to help them walk long distances searching for females. The leg span of males can be more than 2 inches.

- **These spiders** feed on small invertebrates such as beetles, cockroaches, earwigs, and even earthworms.

- **Sometimes house spiders** can survive without food for months.

- **Webs are constructed** in the corners of rooms, under tables or chairs, and in window frames.

- **House spiders build** sheet-shaped webs and wait for prey to become trapped.

▶ Male house spiders may slide into the bath by accident while wandering around in search of a mate. House spiders are not dangerous but belong to the same family as black widow spiders, which are deadly.

During the mating process, the male stays with the female for a few weeks and mates a few times before eventually dying. After the male dies, the female eats him.

The egg sacs are brown and have a hard, papery covering.

Some people find the house spider irritating because it frequently spins new webs in different places and can produce many webs in a short period of time.

279

Orb web spider

Orb web spiders are named after the orb-shaped webs they spin to trap their prey. These spiders are also called orb weavers.

They are the largest web-weaving spiders. The webs vary greatly in size and structural design. Some species decorate the webs with extra silk.

Only the females weave webs. Most build a new web every night. They build their nests in open areas and between flower stems. It usually takes an hour to build a web.

A thin, sticky thread is released to start making the web. This thread is carried by the wind, and when it sticks to something, the spider walks along the thread to make it thicker. This is repeated until a strong web is ready.

These spiders are found in a variety of sizes and colors. They usually have hairy bodies and legs. Their hair helps them to sense any activity around them.

Although most orb web spiders are blind or have very weak eyesight, they can distinguish between day and night.

There are many kinds of orb web spiders. Some of the most common are the golden orb web spider, the St. Andrew's cross spider, and the cross or garden spider.

Male orb web spiders pluck strands of the web to attract the attention of the females. Females are usually larger than the males.

When prey is trapped in the web, the spider wraps it in a sticky, silky substance so that it does not tear the web apart. The wrapped prey is then either eaten immediately or stored and eaten later.

When disturbed, most orb web spiders fall to the ground in alarm. However, this is not true for all species. For instance, when a female St. Andrew's cross spider is disturbed, she grasps her web and shakes it.

▲ *This fly is caught in the web of an orb web spider.*

Spitting spider

Spitting spiders belong to the family of six-eyed spiders, Haplogynae. They have six eyes and long, thin legs.

It is believed that spitting spiders originally came from the tropics. They are now found worldwide. In cooler climates, they are most likely to be found inside buildings.

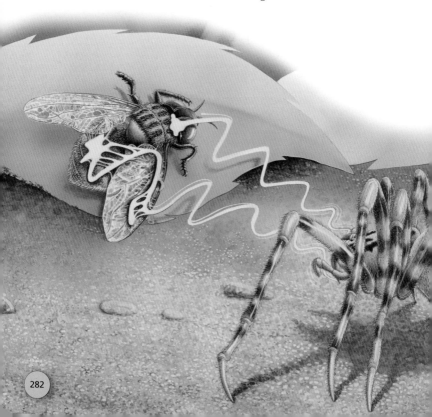

DID YOU KNOW?

Spitting spiders feed on mosquitoes, moths, and flies. They often swing their heads from side to side while spitting, creating a zigzag stream that binds the prey to the ground.

These spiders are shiny and have a light brown body with black spots, and legs with black rings. They are quite small, ranging from 3–6 mm. Spitting spiders have a distinctive shape—a large, dome-shaped head and chest area (cephalothorax).

Found all over the world, they live in dark corners of houses or around window frames and in cupboards. Sometimes they can be found outdoors under leaves and bark.

Spitting spiders usually hunt at night. They eat clothes moths and other pests.

Little is known about the mating habits of these spiders.

After laying her eggs, the female carries them around in her jaws. The eggs hatch about two weeks later.

Spitting spiders have a relatively long life span and live for two to four years.

These spiders are not harmful to human beings.

◄ A spitting spider has silk glands, which are connected to its poison fangs. When it spots its prey, it spits a sticky silken substance over the prey to immobilize it.

Funnel-web spider

- **Funnel-web spiders** are medium-sized and have a glossy brown or black body. Their abdomen is usually dark plum or black in color.

- **These spiders** are named after their habit of making sheetlike webs that funnel into a tubular retreat at one point.

- **The spider** hides in this retreat and rushes out when the web vibrates because prey has been trapped.

- **Moist and cool places** are the preferred habitat of the funnel-web spider. It can be found under rocks and in crevices in wood or tree bark.

- **These spiders** cannot jump or swim. The hair on their bodies helps them to trap air bubbles and stay underwater for a few hours.

- **The male follows** a chemical scent produced by the female to find her burrow. While mating, the male takes care that the female does not bite him.

- **After mating**, the female lays her eggs in a silken egg sac. The spiderlings emerge after approximately three weeks.

- **Spiderlings** take about four years to mature, but they leave their mother and fend for themselves after they have molted twice.

The most deadly funnel-web spider is found in Australia and is called the Sydney funnel-web spider. The venom of the male is five times more poisonous than that of the female, and a victim can die within 15 minutes.

▼ *Funnel-web spiders lay trip lines outside their burrows. These alert the spiders to any passing prey or mate and enables them to react quickly.*

Black widow spider

🦟 **Female black widow spiders** are one of the most feared and dangerous spiders. Only the females can deliver a fatal bite to humans. Males and young black widows are harmless.

🦟 **Easy to recognize**, females are black and have a red hourglass mark on their abdomen.

▼ *After mating, the female black widow sometimes eats the male, which is why she is called a "widow." She stores the sperm in her body so she can produce more eggs without mating again.*

- **Males** are about half the size of females, but their abdomen is more elongated and they have longer legs in proportion to their body. They have red and white markings on their abdomen.

- **Found in the warm** and temperate regions of the world, such as the United States, Italy, and South Africa, these spiders live in dark places such as drainpipes or under rocks and wood. They are not usually found in houses.

- **Black widows eat** flies, moths, and other small insects but their venom also works on larger mammals.

- **Some female black widows** live for more than three years; males only live one or two months.

- **Female black widows** construct an irregular-shaped web, which is tangled. The core of the web is like a funnel and is designed to capture large insects.

- **Once prey** is captured, the female wraps it in silk and kills it by injecting venom. The venom digests the prey and the spider is able to eat it easily.

- **The female** lays her eggs in egg sacs, each containing several hundred eggs. She stays near her eggs and guards them. During this period, the female is more likely to bite if disturbed. At other times, when threatened, these spiders tend to escape rather than bite.

- **After about 20 days**, the spiderlings tear the egg sacs and emerge. However, they tend to be cannibalistic and eat each other.

Brown recluse spider

🕷 **Brown recluse spiders** are small and brown. They are similar in appearance to violin spiders, which have violinlike patches on their body.

🕷 **The brown recluse** spider is named after its reclusive life spent away from others.

🕷 **These spiders** are approximately half an inch long and weave a sheetlike web in dark, secluded corners of people's homes, under rocks, or in forest areas.

🕷 **Their small, irregular webs** can be seen in attics and cellars, as well as behind furniture and stacks of wood.

🕷 **Long legs** covered with tiny hairs give these spiders a velvety appearance.

🕷 **Unlike most spiders**, they have six eyes that are arranged in pairs over their head.

🕷 **Brown recluse spiders** are nocturnal and usually hunt for their food, mainly insects, at night.

🕷 **Unaggressive** creatures, they bite only when threatened by humans. Sometimes they can nest in unused shoes or clothing, and if accidentally touched, they can bite.

DID YOU KNOW?
These spiders are highly venomous. A bite can cause irritation and even long-lasting sores. If people are allergic to the venom, the bite can be life-threatening.

The bite of a brown recluse spider leaves a scar, which heals over a period of time.

Brown recluse spiders can live up to ten years.

Males are smaller than females. After mating, the females lay about 50 eggs.

▼ *Female brown recluse spiders protect their eggs inside silk egg sacs. Spiderlings emerge from the sac in three to five weeks and molt six or seven times as they grow into adults.*

Trapdoor spider

Trapdoor spiders belong to the order Araneae and measure 0.5-1 inch in length.

These spiders make burrows in the ground. They build a silk-hinged door at the opening of their burrows. The spiders wait for prey to pass, and quickly come out of this door to grab a meal.

Trapdoor spiders feed on crickets, moths, beetles, and grasshoppers. They generally live for up to 20 years.

▶ Trapdoor spiders make the doors to their burrows using alternating layers of earth and silk. The top surface of the trapdoor may be disguised by a covering of gravel and sand.

- **Some species** lay traps for their prey by stretching a silk "trip wire" from the burrow. As soon as a prey trips on the silk line, the spider rushes out and bites the victim. The venom kills the prey and also aids digestion.

DID YOU KNOW?
Trapdoor spiders are incredibly strong. They can move objects that weigh up to 140 times their own weight!

- **Trapdoor spiders** in the Ctenizidae families have a special row of "teeth" on their mouthparts, which are adapted for digging.

- **Birds, centipedes, scorpions, and wasps** prey on trapdoor spiders. Spider-hunting wasps flip open the trapdoor and paralyze the spider with their sting.

- **Some species** of trapdoor spiders build burrows with side chambers and extra doors to help them escape from predators.

- **The female** usually spends her entire life inside the burrow and can mistake a male trapdoor spider for prey.

- **A male** is guided by pheromones (chemical scents) produced by the female and also by the pattern of silk around her burrow. Some males perform courtship dances to attract the females.

- **The eggs** are laid inside the burrow and are covered in a cocoon.

- **Not poisonous to humans**, the bite of a trapdoor spider may nevertheless cause some pain and swelling.

Mouse spider

🕷 **Mouse spiders** are so called due to their gray stomachs covered with hair, which looks like mouse fur. These spiders are sometimes mistaken for funnel-web spiders.

🕷 **These spiders** belong to the trapdoor family of spiders. The two main species are the red-headed mouse spider and the eastern mouse spider.

🕷 **Ranging in size** from 0.5–1 inch, these spiders have short, stocky legs and many tiny eyes spread across their head.

🕷 **These spiders** can be found in all kinds of habitats, from deserts to rain forests.

🕷 **Mouse spiders live** in oval burrows, which they dig in the ground. Hidden trapdoors cover these. After covering the walls of the burrows with digestive fluids and mud, the mouse spider lines the walls with silk.

🕷 **Females spend** their entire life in the burrow, but the males wander freely. Mouse spiders feed mainly on insects, but they are also known to eat frogs, lizards, mice, small birds, and other spiders.

🕷 **Mouse spiders** do not spin webs to catch their prey. Instead, they usually hunt for them at night. Once caught, the victim is crushed and then immobilized with venom before the spider feeds.

🕷 **Few males** escape alive after mating. The female lays about 60 eggs, which she wraps in a silk cocoon and places in a side burrow halfway down her own burrow.

Once the spiderlings hatch, they continue to share the burrow with their mother. By doing this, many spiderlings are able to survive.

When they are in danger, male mouse spiders can be aggressive and usually rise up on their hind legs. They have large fangs that can deliver painful and dangerous bites.

▲ *A male red-headed mouse spider leaves its burrow. Females are smaller than males, and they are entirely black.*

Tarantula

Tarantulas are hairy spiders found in the warm regions of the world, such as South America, the southern parts of Asia, and Africa.

There are about 800 different species of tarantulas.

◀ *This massive, hairy Goliath bird-eating spider has a 12-inch leg span. That's bigger than your head!*

- **The hairs** on a tarantula's body are sensitive to touch, taste, and vibrations in the air (sounds). The barbed hairs on the abdomen come off easily and can irritate human skin.

- **Tarantulas usually** dig underground burrows, but some live in burrows dug by rodents or other animals. Some species live on the ground as well as in trees.

- **Tarantulas eat** a variety of animals, from insects and small reptiles to small birds and frogs. This is why they are sometimes called bird-eating spiders.

- **Their powerful jaws** can be used to crush their prey. A tarantula's jaws bite straight downward instead of moving sideways like the jaws of most spiders.

- **Active at night**, tarantulas search for prey or wait outside their burrows for it to come near them. They kill their prey with their poisonous fangs. Then they inject a chemical to dissolve its flesh so that it is easy for them to eat.

- **Male tarantulas** search for a female by following the scent she produces. After performing the courtship dance, they mate. The female then lays eggs in her burrow and weaves a cocoon to protect them.

- **The life span** of these spiders is longer than that of other spiders. Females can live for up to 20 years, but males do not live as long.

- **When cornered**, tarantulas purr and raise their front legs in a defensive position.

Hunting spiders

🦟 **Hunting spiders** include water spiders, wolf spiders, wandering spiders, nursery-web spiders, huntsman spiders, and lynx spiders.

🦟 **Most species** do not construct webs to trap their prey.

🦟 **Some of these spiders**, however, construct large webs on the ground and run after insects that land in the web.

🦟 **Large eyes** enable hunting spiders to see their prey from a considerable distance. However, crab spiders, water spiders, and tarantulas have small eyes.

🦟 **Hunting spiders** have strong chelicerae (a pair of "jaws" used for clasping and killing the prey), which help them to hunt and capture their victims.

🦟 **The cephalothorax** (the fused head and thorax region) is usually larger in hunting spiders than in web-spinning spiders.

🦟 **Some hunting spiders** lie in wait for their prey and pounce on it. Others chase after their prey.

🦟 **These spiders have flattened bodies** and legs that point forward to enable them to slip under loose bark or stones quickly and easily.

🦟 **Vegetation and tree trunks** are the preferred habitats of hunting spiders.

DID YOU KNOW?

Hunting spiders that do not weave webs spin a silken thread and hang from it. They can use these threads to descend to the ground from high places.

▼ Hunting spiders such as this wolf spider have excellent eyesight, which helps them find prey.

Crab spider

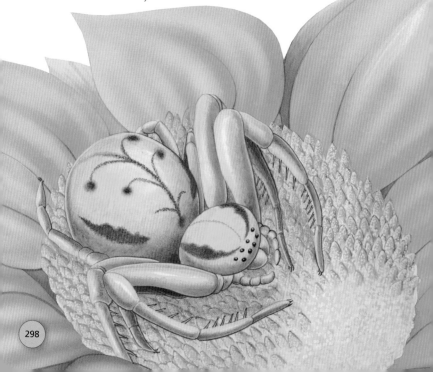

Crab spiders belong to the family Thomisidae of the order Araneae. There are about 2,000 different species of crab spiders.

These spiders are named for their crablike shape and their ability to walk sideways or backward. They are easy to identify, as their first two pairs of legs are much longer than the rest.

Crab spiders have very small eyes. Although their eyesight is not very good, they can sense movement from almost 8 inches away.

🐝 **The female is larger** than the male and her body is almost round.

🐝 **Crab spiders do not spin webs**. They move slowly and are not good hunters.

🐝 **These spiders hide** themselves and lie in wait for their prey. As they can change their color to match their environment, it is difficult for their prey to spot them. This helps them to make surprise attacks.

DID YOU KNOW?

Crab spiders are often found on flowers, which is why they are also known as flower spiders. They can change their color to match that of the flower on which they rest.

🐝 **Some crab spiders** resemble tree bark, leaves, or fruits, while others look like bird droppings.

🐝 **Many crab spiders** use venom to immobilize their prey. This helps them to kill insects much larger than themselves.

🐝 **Crab spiders** do not wrap their prey in silk after capturing and biting it. They remain with it until they have sucked it dry.

🐝 **These spiders prey** mainly on butterflies and bees that come to seek nectar from flowers.

◀ The giant crab spider often hides camouflaged on flowers and attacks insects that come to feed there.

Fishing spiders

🦟 **Raft spiders**, also called fishing spiders, belong to the order Araneae. They are dark brown with stripes on each side of their body.

🦟 **Males measure** 0.5 inch in length, and females are almost twice as big.

🦟 **Raft spiders** live near water and do not spin webs. They feed on tadpoles, small frogs, and insects.

🦟 **To locate its prey**, a raft spider lowers its front legs into the water and feels the vibrations. The spider then pulls its prey out of the water and feeds on it.

🦟 **Raft spiders** can crawl down water plants if threatened, and can remain underwater for 40 minutes or more.

🦟 **During courtship**, the male spider attracts the female by making regular waves on the surface of the water. He jerks his abdomen up and down and waves his legs in the air.

🦟 **The female** is very aggressive toward the male and in some cases she might even eat the courting male.

🦟 **After mating**, the female lays more than 1,000 eggs in a large egg sac.

🦟 **Green in color**, the egg sac is carried around by the female. When the spiderlings are about to emerge, the female spins a protective silk web around the egg sac. She opens the egg sac and guards it until the spiderlings emerge.

▼ A raft spider waits patiently for the right time to strike. Its sensitive legs touch the water, feeling for any vibrations caused by possible prey, such as fish or frogs.

DID YOU KNOW?

The raft spider can walk on the water's surface without sinking. It spreads its legs out wide and takes quick, gentle steps.

Water spider

- **Water spiders** live in freshwater ponds, lakes, and slow streams.

- **These spiders** are dark brown in color. They have hair on their abdomen, which helps them to trap air bubbles and breathe underwater. This gives the spider a silvery appearance.

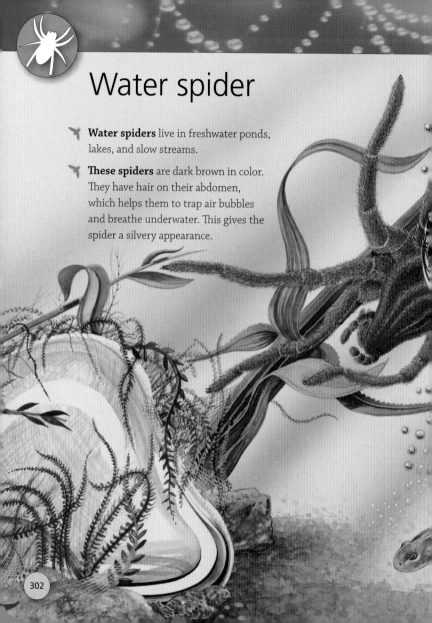

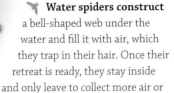

🗡 **Water spiders construct** a bell-shaped web under the water and fill it with air, which they trap in their hair. Once their retreat is ready, they stay inside and only leave to collect more air or to catch food.

🗡 **These spiders** can deliver a painful bite, which can cause irritation to humans.

🗡 **Water spiders** eat small fish and aquatic insects.

🗡 **After mating**, the female water spider lays her eggs in a white egg sac in her "diving bell." The spiderlings emerge in late spring or early summer.

🗡 **While molting**, water spiders usually come out of the water to shed their old skin on vegetation near the pond.

🗡 **The life span** of a water spider is about two years.

🗡 **These spiders** hibernate in winter. In late autumn, the water spider seals its diving bell and stays there all winter.

> **DID YOU KNOW?**
> Water spiders are good swimmers and, once they catch their prey, they clasp it with their venomous jaws and carry it back to their underwater retreat.

◄ Water spiders ferry air from above the water down to their "diving bells," which are anchored at the bed of a stream, river, or pond.

Lynx spider

Lynx spiders are named for their habit of pouncing on their prey like a lynx. They are very swift runners.

Easily recognized by their narrow, conical abdomen, lynx spiders have spines on their body and long, thin legs. The cephalothorax is oval and the fangs are short. Males are smaller than females.

These spiders are usually green in color, but they are sometimes orange-brown with yellowish-green and silver stripes on their body. These colors help to camouflage them among vegetation.

Lynx spiders have fairly good eyesight, and their eight eyes are arranged in a hexagonal pattern.

Commonly found on trees and in vegetation, they do not build webs, but wait on plants for insects and then pounce on them.

Mainly active during the day, they can be seen running over vegetated areas and leaping swiftly from place to place.

If disturbed, these spiders jump away. They seldom bite humans. However, when they do, the bite is painful.

Females do not lay their eggs in webs. Instead, they attach them to plants. They are very protective of their eggs, which hatch after two weeks.

One species of lynx spider, the green lynx spider, is considered very beneficial for agricultural pest management.

▶ *Lynx spiders are hunters, not web builders, and they use speed, agility, and good eyesight to seek and capture prey.*

Wolf spider

Wolf spiders belong to the order Araneae and the family Lycosidae. The meaning of *lycosa* in Greek is "wolf."

These spiders measure up to 1 inch in length. They are usually found in North America, Europe, Asia, and Australia.

Found in grass or under stones, logs, or leaf litter, their habitats range from seashores to mountainous areas. Some wolf spiders can even skip across the surface of the water and duck under it for short periods.

These spiders are named for their wolflike habit of chasing and pouncing on their prey. They are also known as ground spiders or hunting spiders.

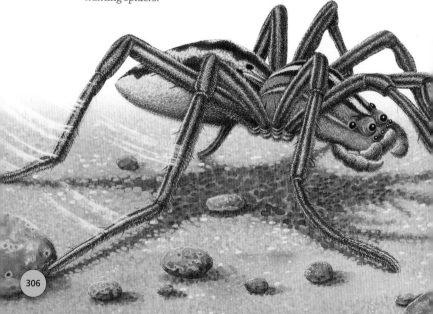

🦟 **Most wolf spiders** are dark brown in color and have a long, broad, hairy body with long, strong legs.

🦟 **Ground dwellers**, wolf spiders are known for their very swift running speed.

🦟 **These spiders have good eyesight**. The number and arrangement of their eyes helps scientists to identify them. They have two small- or medium-sized eyes in the top row, two very large eyes in the middle row, and four small eyes in the lowest row.

🦟 **Wolf spiders** use their front legs to grab prey, then bite to inject venom and crush their prey with powerful mouthparts.

🦟 **Courting wolf spiders** signal to potential mates by waving and drumming their palps (mouthparts) and legs.

🦟 **The female** makes an egg sac of papery silk, which she usually carries around. The sac is shaped like a ball and is attached with strong silk to her spinnerets. After they hatch, the young spiders ride on their mother's back for several days.

◄ *The wolf spider is a hunter and does not usually spin webs. It lies in wait, still and silent, ready for an insect or spider to pass by. When prey, such as a cricket, approaches, the wolf spider pounces.*

307

Jumping spiders

Jumping spiders belong to the family Salticidae of the order Araneae. As their name suggests, they can leap across distances that are many times the length of their body. Hairy tufts on their feet help jumping spiders to grip surfaces.

Although mainly found in tropical regions, they also live in rain forests and on high mountains. These spiders love sunshine, and retreat into their silken nests on cloudy or rainy days.

Jumping spiders are usually less than an inch in length. The female is generally larger than the male.

The spider jumps by means of muscular contractions in its body, which forces body fluids into the legs. This causes the legs to extend rapidly. Some species do not jump but scurry around in an antlike manner.

Many species of jumping spiders are brightly colored and patterned. They have stout bodies, short legs, and large eyes on the front of the face.

▼ When jumping spiders leap, they often produce a tough line of silk, called a dragline, which keeps them secure in case they fall or are blown off course.

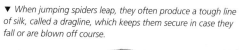

▲ *Most spiders have poor eyesight, but a jumping spider's eyes form clear images of its prey. They can see clearly up to 12 inches away.*

A jumping spider has much sharper vision than other animals of a similar size. It can see in color.

Active predators, jumping spiders usually hunt during the day. They stalk prey and then pounce when they are close.

Most jumping spiders feed on insects, and some feed primarily on web-building spiders. Some even feed on other jumping spiders.

The male's front pair of legs are colored and have distinctive bands of hair. In many species, the male performs a complex courtship dance in which he moves his body and waves his front legs to attract a female.

After mating, the female lays her eggs in a silk-lined shelter under stones or bark, or on the surface of plants. The female often guards the eggs and the newly hatched spiderlings.

Daddy longlegs spiders

▼ *Daddy longlegs spiders feed on insects and other spiders.*

Daddy longlegs spiders belong to the family Pholcidae. They have a short, round cephalothorax with the head region elevated on the sides. The thorax region has a deep, long central groove.

These creatures are very small and have long, thin legs and a long, cylindrical abdomen.

Daddy longlegs spiders are not very active and are usually found in residential areas, houses, buildings, and even caves. They can often be seen in trees and under stones.

The front eyes are small. The other eyes are large and are arranged in groups on either side.

These creatures have venomous glands. However, they do not bite or harm humans. Their fangs are too small to cut or bite the skin.

Daddy longlegs spiders feast on other spiders (including house spiders) and insects. They are natural enemies of the red-back spider.

Sometimes, daddy longlegs spiders invade the nests of insects and eat the prey as well as their eggs.

Their webs are either sheetlike or irregular in shape. The daddy longlegs spider hangs in an inverted position beneath the web.

The female carries her eggs in her jaws in a bundle until the spiderlings hatch.

Scorpions

🗡 **Some scorpions** can kill people with their deadly sting, but most have mild stings that are no more dangerous to people than a wasp sting.

🗡 **Scorpions sting** mainly in self-defense and to paralyze their prey. The venom is not usually poisonous enough to kill humans and other animals.

🗡 **Easily identified** by their venomous tail at the end of an elongated body, they also have four pairs of legs and two lobsterlike claws (pincers) called pedipalps. The average size of a scorpion is 4 inches.

🗡 **Scorpions use their large pincers** as weapons to catch their prey, tear it into small pieces, and crush it into a mushy pulp that can be sucked up by the mouth.

🗡 **These arachnids** can be tan, red, black, or brown in color. A hard exoskeleton protects the scorpion from external damage.

🗡 **Tiny sensory hairs** cover the body and legs and help scorpions to detect temperature changes and movement around them.

🗡 **Scorpions have book lungs**, which are structures for breathing that are similar to gills.

🗡 **Scorpions can store** a large amount of food inside their body. This allows them to live without food for up to a year.

🗡 **In extreme weather** conditions, the extra layer of fat under the scorpion's exoskeleton helps it to survive.

🗡 **At night**, scorpions use the stars to navigate and move around.

▼ Scorpions use their large pincers to grab prey, which may be bigger than themselves.

313

Scorpion life cycle

Scorpions are nocturnal creatures. Once it is dark, they come out of their nesting places, such as crevices, burrows, or under rocks, to hunt for food or find a mate.

Scorpions are known to breed during the late spring, summer, and early autumn.

The male locates a female by following chemical scents called pheromones.

Mating is an elaborate ritual. The male grabs the female and performs dancelike movements before mating.

In some species, the male is killed and eaten by the female once mating is over.

▼ The average life span of a scorpion is around three to five years, but some species are known to live for as long as 25 years.

▲ *In some species of scorpions, the female carries the young on her back for about two weeks until they can fend for themselves.*

After mating, the female retains the eggs in her body and gives birth to live young.

Young scorpions are whitish and are surrounded by a membrane at the time of birth.

On average, a female produces 25–35 young at a time, and she protects them until their first molt.

Young scorpions usually molt four to nine times before becoming adults.

315

Giant hairy scorpion

- **Giant hairy scorpions** are among the largest scorpions in North America and have hair all over their bodies.

- **These scorpions** have brown bodies and yellow legs, and can be easily recognized.

- **Although** they have very weak eyesight, the hairs on their bodies help them to detect ground and air vibrations and any other movement near them.

- **Giant hairy scorpions** have a long tail, with a bulblike poison gland at the tip. They have large pincers called pedipalps to clasp their prey.

- **Aggressive creatures**, they sting with the slightest provocation. Their sting is painful but not fatal.

- **Giant hairy scorpions** are carnivores and feed on small insects and baby lizards.

- **Mostly found** in desert regions, such as in California and Arizona, these scorpions can withstand extremely hot weather and are commonly found under rocks. They also burrow into the ground.

- **Giant hairy scorpions** are nocturnal. During the day, they rest to avoid the heat. At night, they hunt for food or find a mate. They lie in wait to ambush their prey, then grab it with their pincers.

These scorpions maintain the moisture content in their body by sucking the fluids from their prey.

Some people keep giant hairy scorpions as pets. They feed them crickets.

▶ Also known as desert hairy scorpions, the giant hairy scorpion prefers dry, hot habitats. It can reach up to 6 inches in length and is able to hunt and catch prey larger than itself.

Whip scorpion

Whip scorpions are named after the long, whiplike tail at the end of the abdomen. Their first pair of legs is very long and thin and ends in antennaelike filaments, which are used as sensory feelers.

There are over 100 species of whip scorpions worldwide. The length of these scorpions varies from 1–2.75 inches. The largest species of whip scorpion is called *Mastigoproctus giganteus* and can measure up to more than 3 inches in length.

▼ Whip scorpions are arachnids, but they are not considered true scorpions. They are usually larger than scorpions, and their whiplike tails do not contain stingers.

- **Commonly found** in tropical and subtropical areas, they usually live in burrows underground.

- **Whip scorpions have a flat body** and can crawl into cracks and crevices. These scorpions can also be found under logs, rotting wood, and rocks. They prefer to live in moist, dark places.

- **Quite harmless**, whip scorpions have no venom glands and do not bite or sting.

- **These scorpions have glands** situated near the rear of the abdomen. When disturbed, these glands secrete a spray, which is a combination of formic acid and acetic acid. This spray smells like vinegar because of the acetic acid, which is why whip scorpions are also called vinegarones.

- **Whip scorpions are carnivorous** and feed on insects and worms. Some large species of whip scorpions also attack frogs.

- **While mating**, the male transfers a sperm sac to the female. The female lays her eggs in a burrow and covers them with a slimy mucous membrane to keep them from drying out.

- **After the eggs hatch**, the young climb onto their mother's back and cling on with the help of special suckers. They molt three times in about three years before maturing into adults.

- **After the first molt**, the young scorpions resemble the adults and leave the burrow. Soon afterward, the mother scorpion dies.

Pseudoscorpion

Pseudoscorpions are not scorpions. They look like scorpions but do not have the long tail with the stinger at the tip. There are more than 3,300 different species of pseudoscorpions.

These tiny creatures range between 1–8 mm in length. Like scorpions, pseudoscorpions have two lobsterlike pincers with which to grab their food.

Yellow or brown in color, pseudoscorpions have oval bodies. They are popularly known as false scorpions or book scorpions.

Found almost everywhere, they live in forests, leaf litter, grass, birds' nests, and anthills. They can also be found under rocks and on the bodies of insects and birds.

The pincers are usually small, reaching up to 3 mm in length. Many also contain a poison gland. This poison is used to paralyze their prey, such as springtails, mites, booklice, and other tiny invertebrates.

Pseudoscorpions cannot chew. To digest food, they first pour digestive fluids (secreted by their mouthparts) on their prey and then they suck up the liquid food.

Pseudoscorpions have silk glands, which have openings on the jaws, or chelicerae. They use the silk to spin cocoons, in which they molt and live in during winter.

The male performs an elaborate courtship dance before mating with the female.

After mating, the female carries the eggs around in a pouch attached to her abdomen. The eggs hatch into young and undergo two to three molts as they develop into adults. Adult pseudoscorpions live for two to three years.

▼ Pseudoscorpions produce venom in their pincerlike pedipalps, which they inject into their prey, such as this springtail.

DID YOU KNOW?

Pseudoscorpions often attach themselves to the legs of insects such as houseflies, crane flies, and beetles, to hitch rides. They can also walk backward.

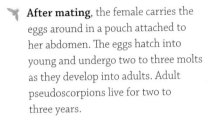

Horseshoe crab

Horseshoe crabs are the only marine arachnids in the world. They are also called king crabs.

Covered by a horseshoe-shaped shell, the cephalothorax consists of the head, neck, and chest. The segmented stomach is covered by a protective shield. The rear end of this shield has a stiff, tapering tail, or telson.

▼ *A horseshoe crab's body is protected by its tough exoskeleton and its drab brown color helps to camouflage it against sand and mud. Horseshoe crabs have two pairs of eyes.*

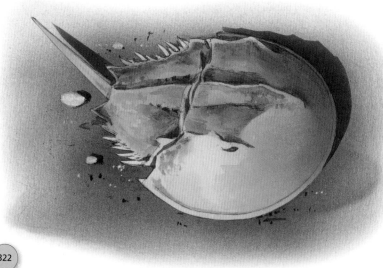

Since this arachnid has book gills for breathing and a shield studded with horns, naturalists mistakenly called it a king crab when it was first discovered. In reality, the king crab, or horseshoe crab, is related to spiders and scorpions, not crabs.

The underside of a horseshoe crab has a pair of small, pincerlike jaws and a mouth surrounded by five pairs of legs, which are used for walking and eating.

Horseshoe crabs are found along the coasts of Asia and North America. They usually live on the seabed in moderately deep waters. In temperate countries, they breed in spring and appear in shallow waters to lay their eggs.

These marine arachnids are active at night. During the day, they live partly buried in mud or sand.

Horseshoe crabs feed on sea worms and other sea creatures, which they find while burrowing. Worms and shellfish are broken into bits before being eaten.

Females are larger than the males and measure about 20 inches in length. Although they look alarming, these arachnids are harmless to humans.

Female horseshoe crabs lay about 20,000 eggs in one night. When the young hatch out, they look like tiny adults, but do not have a tail. It takes them eight to ten years to develop into adults. Horseshoe crabs can live for 20 years or more.

Like starfish, horseshoe crabs can grow new legs to replace legs damaged by predators or the environment.

Ticks

Ticks are invertebrates that closely resemble their relatives, the mites, and belong to the same arachnid group as spiders and scorpions. They can be found all over the world. Around 850 species of ticks are known to exist.

Parasitic, they feed on the blood of birds, some reptiles, and mammals such as humans, dogs, and cows.

Ticks range between 2.5–11 mm in length after feeding. They prefer to live in temperate regions, in habitats such as grasslands.

They have four pairs of clawed legs. They cannot run, fly, or hop like other arachnids. However, they climb on grass and small plants, as well as on man-made structures such as walls.

Ticks have strong sensory organs, which help them to sense the presence of a host so they can be ready to attack.

They climb onto the host and grasp with their legs. Immediately, they sink their mouthparts into the host's skin and start sucking its blood. When they have had enough food, they drop away from the host.

Mating occurs on the host's body. After mating, the female tick drops to the ground and lays her eggs. Male ticks usually die after mating.

There are two major groups of ticks—hard and soft ticks. In a hard tick, there is a hard, platelike shield covering the back and the mouth is visible from above.

Soft ticks have a spongy, wrinkled back and their mouthparts are hidden under their body. They mainly feed on the blood of birds.

The bite of a tick causes constant itching, which can persist for months. Ticks also transmit diseases to humans. The deer tick can transmit lyme disease, which causes fever, rashes, and swollen joints.

▼ A tick is about the size of a rice grain. It has eight legs and is an arachnid, not an insect. When it has just gorged itself on blood, its baglike body swells up like a balloon.

Mites

🐦 **Mites are oval-shaped creatures** that belong to the order Acarina of the class Arachnida. They are found all over the world and around 50,000 species are known to exist.

🐦 **Mites have an unsegmented body**, which means that the head, thorax, and abdomen are fused together. It has four pairs of legs. However, in the larvae stage it has only three pairs of legs.

🐦 **These insects** have adapted well to both land and water habitats.

🐦 **Mites breathe through tracheae**. These are small, tubelike structures that open onto the surface of their body.

▶ These tiny creatures are so small that they may be invisible to the naked eye, but they are nevertheless an important group of arachnids. Some mites are free-living, others are parasites.

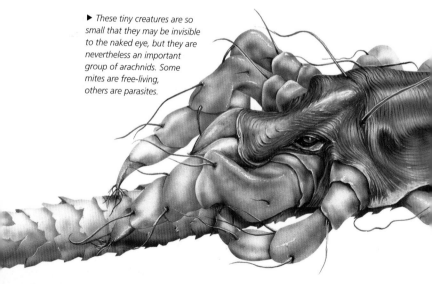

- **Mites may live** freely in the soil or water, but a large number are parasites of plants or animals.

- **The most dangerous pests** for humans and livestock such as cattle, they carry and transmit serious diseases.

- **Follicle mites** infect human hair roots. Bird mites harm the skin of birds. Chicken mites are known for spreading diseases in poultry.

- **Scabies mites** attack horses, dogs, and rabbits and cause itching and bleeding. Some Australian species can also inflict lethal bites.

- **Mites can also live** on vegetation and plant fluids. They cause formations like galls on leaves. Some common plant mites are red spider mites. They spin a web on the undersides of leaves.

Red velvet mite

Red velvet mites are soft-bodied. They are usually red or orange in color and are 0.5–4 mm in length.

Their entire body is covered with a fine coat of hair, making it look like velvet.

Red velvet mites have four pairs of legs. Their front legs are used as organs to sense their surroundings.

The larvae of red velvet mites are external parasites of insects such as locusts, crickets, and grasshoppers. They feed on the blood of these insects and some cling to their wings, even in flight.

Adult red velvet mites are predators of insect eggs, larvae, and pupae and also feed on other small arthropods.

Red velvet mites live in gardens, deserts, and forests, where they are often found walking slowly over the forest floor.

After mating, the female lays her eggs in the soil.

The eggs hatch quickly and the larvae search for an insect host.

Once they have sucked enough blood from their host, the larvae detach themselves and make a burrow in the soil, where they molt and emerge as adults.

Velvet mites are becoming resistant to pesticides. Therefore, people are developing other ways to control them.

▼ Red velvet mites are eye-catching arachnids that live in forests and gardens. Their beautiful color warns predators that they taste particularly unpleasant.

Harvestmen

Harvestmen (sometimes called daddy longlegs) may be mistaken for spiders. They are not spiders, and belong to the order Opiliones.

Unlike spiders, harvestmen do not have spinnerets for secreting silk thread to make nests and webs.

▼ *Omnivorous insects, harvestmen feed on small insects, insect larvae, and spiders. Some are scavengers, feeding on dead creatures as well as animal dung and droppings.*

- **At first glance**, spiders and harvestmen appear similar. Whereas spiders have two distinct body parts separated by a waist or stalk, harvestmen have a single, rounded or oval body shape.

- **These insects are called harvestmen** because in Europe large numbers of them appear in autumn, which is the harvest season.

- **Harvestmen are usually found** in temperate regions of the Northern Hemisphere and in Southeast Asia. About 6,400 species of harvestmen are known to exist.

- **These insects can be spotted** living in hedges, parks, gardens, or anywhere near vegetation. They usually gather in large numbers.

- **Usually 4–10 mm long**, harvestmen have fairly small, oval bodies. They usually have extremely long, thin legs. Many species do, however, have short, thick legs. Certain species are able to shed their legs when they need to escape from their enemies. In some species, the last segment of the leg has many joints, giving it extreme flexibility.

- **Harvestmen have two eyes** in the middle of their head. They appear to be looking sideways.

- **As nocturnal creatures**, they are most active at night. During the day, they hide away from the light by resting under hedgerows or in crevices.

- **Harvestmen are harmless** and do not have poison glands. They cannot sting.

Worms, millipedes, and centipedes

Worms

Worms are long, soft-bodied animals without legs. There are probably more than a million different species of worms. They are an important link in many food chains, helping to maintain the ecological balance. However, they are sometimes harmful to other animals and the environment.

The worm world is divided into several groups. The three most important are roundworms, flatworms, and segmented worms.

As cold-blooded creatures, they depend on the environment for maintaining their body temperature. Most species are parasitic, living inside or outside the bodies of other organisms and taking food from them. These other organisms are called their hosts.

Parasitic worms that affect humans are roundworms, tapeworms, whipworms, hookworms, and pinworms.

Certain species of worms are free-living. They move around in the environment and forage for food. Worms can live in any kind of habitat.

In some species, the females emit a bright luminescence (glow) to attract the males for mating.

Aggressive behavior is common in some species. They guard and defend their nests and tubes against other worms of the same species.

Some worms live in a symbiotic relationship with other organisms. A number of scale worms live in the mantle or the cavity covering of mollusks (shellfish) and starfish.

🪶 **Marine worms** filter food particles from the water and clean it up in the process. They are a type of filter-feeder.

🪶 **It is difficult to preserve** most worms because of their soft bodies and the absence of a skeleton. They are rarely found preserved as fossils.

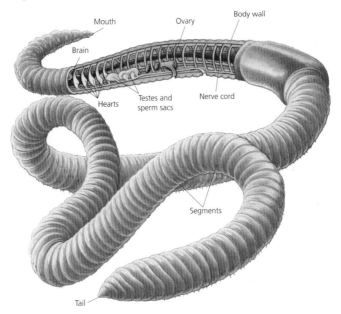

Mouth

Brain

Ovary

Body wall

Hearts

Testes and sperm sacs

Nerve cord

Segments

Tail

▲ *An earthworm is a segmented worm, or annelid, because its body is divided into rings, or segments. The word* anulus *means "ring" in Latin. Tiny bristles on the segments help the earthworm grip the soil as it burrows.*

Roundworms

Also called nematodes, roundworms belong to the phylum Nematoda. These worms are found everywhere in the world. They can live on land as well as in water—both freshwater and salt water.

Roundworms have a round (cylindrical), tapering, and threadlike body. It is not segmented. Most are less than 3 mm long.

These worms have a simple body structure and do not have a well-defined head.

Roundworms have a tubelike digestive system and a simple nervous system. They do not have a circulatory or respiratory system.

There are many types of roundworms, such as eelworms, hookworms, pinworms, and whipworms.

Roundworms have an outer body covering called a cuticle. They usually molt about four times as they grow into adults.

◄ Roundworms are among the most numerous animals. They can destroy farmers' crops.

The largest roundworms are parasites of whales and grow up to 30 feet long.

In humans, a roundworm infection can cause nausea, vomiting, and diarrhea. Doctors prescribe medicine to get rid of the disease. The best way to prevent infections is to eat well-prepared, clean food and take care of personal hygiene.

A roundworm's eggs hatch into larvae inside its host, which can be humans or other animals. These larvae are carried to the other parts of the body through the bloodstream.

DID YOU KNOW?

Roundworms cause dangerous, almost fatal, diseases in humans as well as in other animals. They usually enter the host's body through contaminated food, water, cuts, or wounds.

Lugworm

🦟 **Lugworms have a ringed or segmented body**. The head and gills are dark red. Most of the body is thicker and lighter in color than the head. It can be yellow, black-green, or nearly black.

🦟 **The body** is thin at the tail end, and the middle has bristles. Lugworms can have up to 13 pairs of feathery gills.

🦟 **These creatures** are found in estuaries and on beaches. Their presence is usually revealed by small holes in the sand.

🦟 **Lugworms can be found** in the sand within a U-shaped burrow. One end of the burrow is covered by a funnel-shaped hollow of sand or mud, while the other end is open.

▼ *Large numbers of lugworms live under the sand along the shoreline. The waste material they push out of their burrows piles up on the beach like squiggly heaps of "sand spaghetti."*

Lugworm burrows can be as deep as 2 feet. It takes about two to six minutes for them to burrow down, depending on the amount of water in the sand. The wetter the sand, the less time it takes.

When it wants to feed, the lugworm pumps water into its burrow. The water that flows through its gills contains sand. The worm digests the organic matter from the water and ejects the undigested sand in the form of casts on the surface.

▲ Found deeper in the ground during the winter, lugworms live closer to the surface during the summer. They can stay in their burrow for eight months at a stretch and can survive for as long as one month in frozen sand.

Lugworms are hermaphrodites. This means that the reproductive organs of both sexes are found in the same individual. The eggs of one lugworm are, however, fertilized by the sperm of a different lugworm.

Lugworms play an important part in the ecology of the seashore. They are a major source of food for many fish and seabirds.

These worms are often used as bait by fishermen and are preyed on by flatfish and wading birds such as curlews and godwits.

339

Bloodworm

🕊 **Bloodworms** are marine segmented worms, which belong to the annelid group. They are mostly found on the eastern coast of the United States.

🕊 **The length** of a bloodworm can reach up to 15 inches. Their rate of growth is affected by the presence of food and the temperature and salinity (saltiness) of the surrounding water.

🕊 **Bloodworms** live in tubelike structures built on rocks underwater.

🕊 **Excellent burrowers**, they burrow into sandy mud or silty clay.

🕊 **On their segments**, bloodworms have small, fleshy projections or outgrowths called parapodia, which help them move.

🕊 **The parapodia also contain gills**, which the worms use to exchange gases between their body fluids and the water surrounding them.

🕊 **Bloodworms** have strong jaws containing copper as well as poison glands, which they use to kill their prey. Their poisonous bite is painful to humans.

DID YOU KNOW?
To avoid overcrowding during the winter months, bloodworms redistribute themselves by swimming to another area.

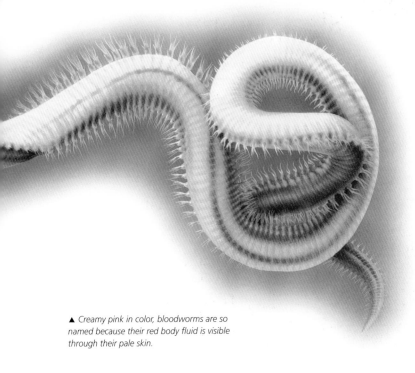

▲ *Creamy pink in color, bloodworms are so named because their red body fluid is visible through their pale skin.*

Predators of bloodworms include other carnivorous worms as well as crustaceans, fish, and seagulls.

After mating, the female releases her eggs into the water. Once the eggs are released, the female dies.

Leech

▼ There are around 650 species of leeches, and they belong to the same group of annelid worms as the common earthworm. Leeches have a sucker at each end of their body—one for feeding and one for hanging on while they feed.

Usually brown or black in color, leeches can range from 1-10 inches in length. They are known for their habit of sucking blood. In some cases, they can even cause death.

Leeches feed on both dead and decaying plant matter. Some feed on insects and other worms, such as earthworms, while other species suck the blood of mammals.

- **Leeches suck blood** by cutting through the skin with their sharp teeth. They fall away from a victim once they have sucked enough blood.

- **Living both on land and in water**, leeches are found in mud pools and lakes, and beneath rocks, leaves, and logs. Aquatic leeches feed on the blood of fish and other sea creatures.

- **Leeches breathe** through their skin. They do not have a digestive system. Instead, they have a pouchlike structure where food is stored.

- **These creatures** lay their eggs in a cocoon, which they may deposit on land or in water.

- **Leeches that live on land** wait for their victims in damp areas, such as rice fields. Once the victim approaches, the leech attaches itself to the victim's skin and sucks their blood.

- **With the help of their victim**, leeches move from one place to another. They attach themselves to the legs or organs of the animals they feed on.

- **About 2,500 years ago**, the ancient Egyptians used leeches to treat all kinds of ailments. Leeches were used until the 1900s to drain the "vapors and humors" believed to cause illness.

- **Today, some species of leeches** are still used to help skin grafts to heal, fight blood clots, and restore normal blood flow during plastic surgery and reconstructive surgery. Leeches produce an anticlotting substance that keeps blood flowing freely while they feed.

Flatworm

🦟 **Flatworms are soft-bodied** worms with flat, ribbon-shaped bodies.

🦟 **The three main types** are tapeworms, flukes, and planarians. Tapeworms live in the intestines of animals. Flukes live in various parts of an animal's body. Both are parasitic. Planarians are nonparasitic, or free-living.

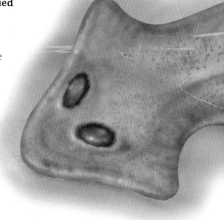

🦟 **Free-living flatworms** are found on land as well as in freshwater and salt water, while parasitic flatworms live inside their host.

🦟 **The outer surface** of free-living flatworms secretes a fluid that forms a hard outer coating called a cuticle. They are covered with short hairs called cilia.

🦟 **Free-living flatworms** are colored, while the parasitic ones are usually colorless.

🦟 **Parasitic flatworms** have a complicated life cycle. They may live in four or five hosts before they complete their life cycle.

🦟 **The reproductive system** of these creatures is complex and occupies a large area inside their bodies.

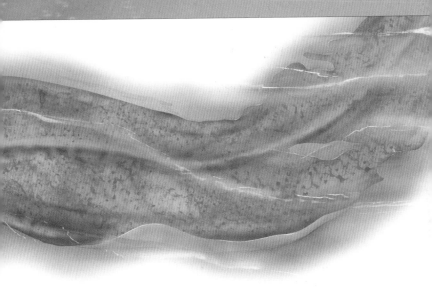

▲ Flatworms are the simplest of all the worm species. They do not have blood and a circulation system. Instead, oxygen and carbon dioxide pass directly through their skin.

All flatworms are hermaphrodites. This means that the male and female reproductive organs are found in the same individual but the eggs and the sperm are formed separately.

An interesting characteristic of flatworms is that they can reproduce asexually. They do this by separating into two halves to produce an entirely new worm. A new flatworm can also develop from a piece of worm that has been cut off.

Well-developed muscles enable these worms to change the shape of their body.

345

Giant tube worm

- **Giant tube worms** are found in the deep ocean near hydrothermal vents, which are hot springs on the ocean floor.

- **These worms grow** up to 10 feet in length. They are firmly attached to the ocean floor and grow upright.

- **Giant tube worms** live inside white tubes made of a hard, tough material called chitin. They have delicate red plumes, which contain blood rich in hemoglobin.

- **Giant tube worms do not have eyes**, a mouth, or a digestive system. At first, scientists found it difficult to explain the presence of blood and how tube worms fed and survived in the dark ocean depths.

- **It was discovered** that giant tube worms have a mouth and a digestive system in their early stage of development. Bacteria enter the tube worms when they are young and remain trapped inside them.

- **It is the bacteria** that feed the giant tube worms. The red plumes trap the oxygen and chemicals needed by the bacteria to manufacture food.

- **Chemosynthesis** is the process of making food without sunlight.

- **The hemoglobin** present in the blood of giant tube worms collects all the chemicals that are needed by the bacteria. These hemoglobin molecules are 30 times larger than those found in humans. The blood carries the required chemicals to the bacteria, which convert them into food.

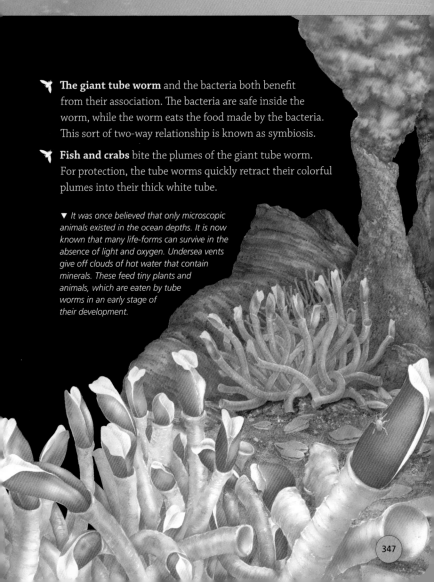

The giant tube worm and the bacteria both benefit from their association. The bacteria are safe inside the worm, while the worm eats the food made by the bacteria. This sort of two-way relationship is known as symbiosis.

Fish and crabs bite the plumes of the giant tube worm. For protection, the tube worms quickly retract their colorful plumes into their thick white tube.

▼ It was once believed that only microscopic animals existed in the ocean depths. It is now known that many life-forms can survive in the absence of light and oxygen. Undersea vents give off clouds of hot water that contain minerals. These feed tiny plants and animals, which are eaten by tube worms in an early stage of their development.

347

Pinworm

🐦 **Pinworms are small**, thin roundworms (nematodes). They are also known as threadworms and are common parasites of humans.

🐦 **These worms** are creamy white in color. Pinworms have three lips on their head. They have pointed tails, which resemble pins.

▼ Pinworms are tiny nematode worms that mainly live in the last section of the host's gut, the large intestine or colon. They are a type of roundworm with longitudinal muscles (arranged lengthwise) rather than circular muscles. This causes them to thrash and curl their bodies when they move around.

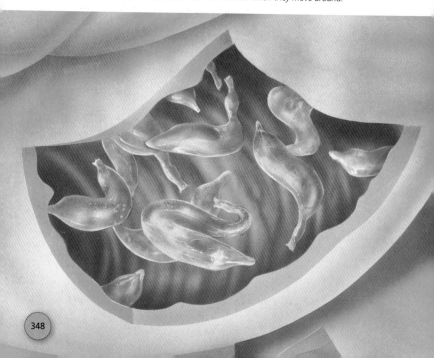

- **Females are much longer** than males. Males have a curved tail, unlike the pointed tail of the female worm.

- **Pinworm eggs** can end up on anything that a person with pinworms touches. They can live for about two weeks and are microscopic.

- **A pinworm egg** enters the digestive system. By the time it travels to the large intestine, the egg has grown into an adult worm.

- **The adult pinworm** attaches itself to the walls of the large intestine. It comes out of the human body to lay eggs at the anus.

- **Then the worm** travels back into the large intestine, while the eggs remain at the anus until they hatch.

- **The life cycle** of a pinworm may take one to two months. The affected person may not be aware of the infection until two months have passed by.

- **Children who are affected by pinworms** may lose their appetite. They do not sleep well at night and feel itchy. They may be restless in the daytime as well. Some people with pinworms may have no symptoms at all.

- **There are many medicines** available to treat the infection caused by pinworms. The best way to avoid getting pinworms is to wash your hands before eating, after playing outside, and after using the toilet. Keeping the fingernails short and clean is also a good idea.

Guinea worm

🦟 **Also known as dragon worms**, guinea worms are common parasites that live in the bodies of humans as well as other animals. The disease caused by this worm is known as dracunculiasis.

🦟 **Guinea worms** can be found in Asia, the Middle East, Arabia, and northern and equatorial Africa.

🦟 **Humans** usually get dracunculiasis by drinking stagnant water that contains tiny water fleas known as copepods that are infected with guinea worm larvae.

🦟 **This parasite** has been associated with people from ancient times. Greek, Roman, and Arabian scholars have mentioned these worms in their writings.

🦟 **References** to a "fiery serpent" that caused havoc among the Israelites may have been the guinea worm. It is also thought that the serpents shown in medical symbols are actually these worms.

🦟 **A female guinea worm** can measure up to 3 feet in length and about 0.08 inch in diameter. The male is about 0.5-1 inch in length.

🦟 **Females** live for 10–14 months, while the males usually die after mating.

🦟 **Both male and female** guinea worms live in the connective tissue of various organs of the human body.

🦟 **Females live just under the skin** of the human host, usually in the legs, ankles, or feet.

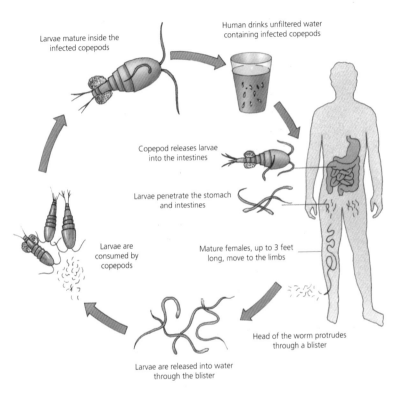

Larvae mature inside the
infected copepods

Human drinks unfiltered water
containing infected copepods

Copepod releases larvae
into the intestines

Larvae penetrate the stomach
and intestines

Larvae are
consumed by
copepods

Mature females, up to 3 feet
long, move to the limbs

Larvae are released into water
through the blister

Head of the worm protrudes
through a blister

▲ *The diagram above shows the life cycle of a guinea worm.*

The developing embryos within the female cause her stomach
walls to burst, releasing the young into the host's body. They can
be found just below the host's skin. Eventually they emerge from a
blister on the body of the host.

Earthworm

🪶 **An earthworm's body** is divided into ringlike segments. Most earthworms have a soft body. They are brown to red in color.

🪶 **Earthworms** have five pairs of hearts. The brain, hearts, and breathing organs are located in its first few segments.

🪶 **Although they cannot see or hear**, earthworms are sensitive to light and vibrations. Their mouth has a flap that helps them to sense vibrations around them.

🪶 **These worms feed** on decaying organisms. They also consume large amounts of soil, sand, and tiny pebbles. They ingest and discard food and soil equivalent to their own weight every day.

▼ Many animals, including the European mole, prey on earthworms.

DID YOU KNOW?

The giant Gippsland earthworm from Australia measures up to 13 feet long.

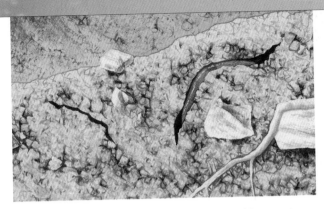

▲ *The earthworm's burrowing activities let air into the soil and decompose organic matter, which helps plants to grow. The earthworm's waste material is also very useful in making the soil fertile.*

Earthworms are hermaphrodites, which means that each worm has both male and female reproductive organs. However, the sperm of one earthworm fertilizes the eggs of another earthworm.

While mating, two earthworms are bound together by sticky mucus and transfer their sperm to each other. The worms separate and form cocoons. Within a day after mating, the cocoon is deposited in the soil. Miniature earthworms usually emerge from the cocoon after two to four weeks.

Earthworms are known as clitellates. They are characterized by a clitellum, or thick band of skin, that produces the cocoon into which eggs are laid and also creates the sticky mucus that covers the worm and helps it to slide through the soil.

Usually found near the surface of the soil, they also make tunnels as deep as 6 feet in winter.

Christmas-tree worm

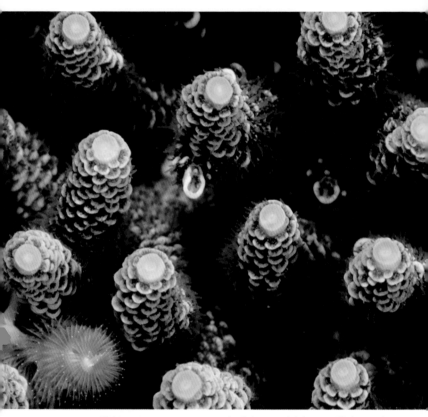

▲ Christmas-tree worms are a type of polychaete, or bristle worm. Their tentacles are intricate, delicate, and often beautifully colored, making them a favorite find of undersea divers and photographers.

- **Christmas-tree worms** are conical in shape and resemble a small Christmas tree.

- **These worms** are found on corals (colonies of tiny animals) in shallow seas. They can be many different colors—yellow, pink, white, blue, gray, and sometimes brown.

- **Corals provide shelter** to many kinds of invertebrates (animals without a backbone), including the Christmas-tree worm.

- **A Christmas-tree worm** builds a tube on the surface of a coral. As the coral grows, the worm grows at the same rate to stay on the surface of the coral.

- **These worms** are excellent at sensing any movement. The worm lies in the tube so that only its head protrudes out of the end. If the worm senses danger, it immediately pulls its head down into its tube.

- **A Christmas-tree worm** is covered in feathery gills that corkscrew around its body.

- **These worms feed on plankton**. The gills are covered in sticky mucus and catch tiny plankton, which float around with the ocean currents.

- **The gills** also allow the worm to breathe by exchanging oxygen and carbon dioxide with the water.

- **Many species of fish**, such as butterfly fish, feed on Christmas-tree worms.

Myriapod

Myriapoda **means** "many-footed," and some of the creatures that belong to this group have up to 400 pairs of legs. Myriapods are arthropods and their bodies are divided into segments. They measure up to 12 inches in length.

Myriapods are a little-known group, although 11,000 species have been discovered. They are nocturnal creatures that have not evolved much over the years.

▲ *A centipede. Centipedes and millipedes are the best-known myriapods. Unlike insects, myriapods do not have a waterproof cuticle or skin and must remain in a humid habitat to avoid drying out.*

There are four groups of myriapods—Symphyla, Pauropoda, Chilopoda (true centipedes), and Diplopoda (millipedes).

Found in soil, leaf litter, or under stones and wood, myriapods are difficult to spot. They prefer humid environments.

Certain species inflict painful bites on humans. The growth of myriapods is anamorphic—they add segments and legs with successive molts. Reproduction is usually by indirect sperm transfer.

Myriapods primarily feed on decaying vegetation. They break down dead vegetable material and play an important role in the ecological balance of forests. Some species are carnivorous and predatory.

These creatures cannot migrate and cannot survive in seawater.

Many species possess repugnatorial glands—glands that secrete foul-tasting chemicals that help to deter predators.

Centipedes have only one pair of legs per body segment and are predatory. Unlike centipedes, most millipedes feed on decaying vegetation, although some are carnivorous.

It is believed that myriapods may have been living on earth as early as 400 million years ago.

▼ Snake millipedes have shiny, cylinder-shaped bodies. They live in leaf litter and climb trees and fences to feed at night.

Centipedes

🪶 **Centipedes have a long**, segmented, wormlike body. Each segment except the last one has one pair of legs.

🪶 **Centipedes are related** to millipedes. They differ from millipedes in having flattened bodies and only one pair of legs on each body segment (millipedes usually have two pairs of legs per segment).

🪶 **These creatures** move very rapidly, as they have anywhere from 14 to 177 pairs of legs. They have one pair of long, many-jointed antennae.

🪶 **Humid environments** are the preferred habitat of centipedes. They are nocturnal creatures and generally remain under stones, bark, and leaf litter during the day. At night, they hunt and capture other small creatures.

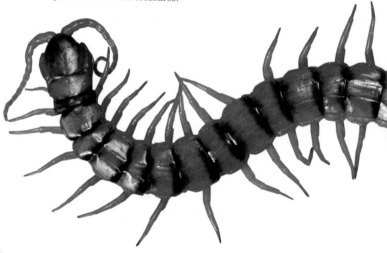

Centipedes are predatory. The first pair of appendages on their trunk is modified into a pair of jawlike claws, which have poison glands. Centipedes use the poison to paralyze their prey.

Centipedes prey on red worms, insect larvae, newly hatched earthworms, and various insects and spiders.

Most species of centipedes are beneficial. However, the bite of large centipedes can cause humans some pain and discomfort, although there are no authenticated cases of adult human deaths caused by centipede bites.

Besides poison, some species of centipedes secrete substances from glands found along their body segments. These secretions are not usually toxic to humans.

Not many fossil records of centipedes have been found because they have a thin and light cuticle, which is not easily preserved.

◄ Centipedes are fierce predators. Their many legs make them swift runners.

Giant centipede

Giant centipedes are so called because they can be as long as 12 inches. They have short, hooklike legs.

They are found in tropical and subtropical regions, both in forests and deserts.

The color of a giant centipede depends on the area in which it lives. Giant centipedes found in rocky areas are gray in color, while those found under the ground are almost black.

While some giant centipedes are colored to blend in with their surroundings, others display markings in vivid yellow and black. These colors alert potential predators to the poisonous nature of the centipede.

During the day, these centipedes rest under stones, wood, and debris.

At night, they emerge to hunt for prey. These centipedes have a pair of long antennae, which helps them to sense the presence of any prey that are nearby.

Giant centipedes feed on insects, scorpions, and sometimes even animals such as small lizards, small birds, and mice.

Giant centipedes paralyze their prey by injecting it with poison, and then they crush it with their mandibles (mouthparts).

After the female has laid her eggs, she wraps herself around them for protection. It usually takes about three weeks for the eggs to hatch.

Giant centipedes can deliver a painful bite. However, they do not usually bite humans unless they are provoked.

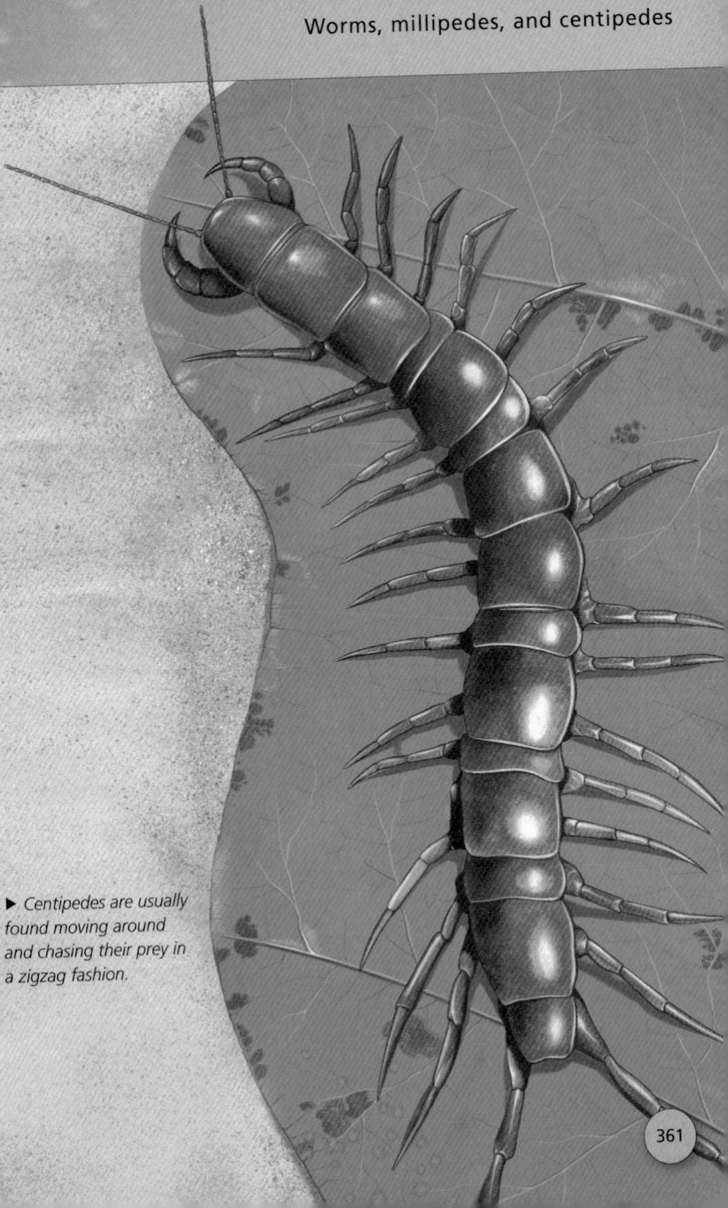

▶ Centipedes are usually
found moving around
and chasing their prey in
a zigzag fashion.

Millipede

Although the word *millipede* literally means "thousand legs," millipedes do not have that many legs.

They belong to the class Myriapoda and are closely related to centipedes. They differ from centipedes in that they lack venom and move more slowly.

Millipedes have a segmented body, a hard exoskeleton, and many jointed legs. Their body is a shaped like a tube.

Depending on the species, they usually have between 45 and 400 pairs of legs.

Most body segments have two pairs of legs. The first few segments, which have only one pair of legs, are called somites, while segments with two pairs of legs are called diplosomites.

▼ Approximately 10,000 species of millipedes have been identified. These many-legged myriapods are similar to their centipede cousins in lifestyle and habitat, but they are slower-moving.

▶ *The rusty millipede is found in most places in the world and measures about 3 inches in length.*

The **penultimate segment** has no legs and the last segment—the anal segment—is used for excreting waste from the body.

Millipedes live in moist places on land under rocks, leaf litter, or logs. Some millipedes are garden pests.

Spiracles, which are holes found on their body, are used for breathing.

Millipedes hatch from eggs. When the young hatch, they have only the first three pairs of legs. Each time they molt, they add more segments and legs to their body.

Most species of millipedes are herbivorous and feed on dead or decaying plant material. A few species are carnivorous.

Birds and shrews prey on millipedes. When they are in danger, millipedes curl up in a spiral to protect their soft underside.

DID YOU KNOW?
When threatened, millipedes can eject a foul-smelling chemical, which repels many predators.

Pill millipede

- **Pill millipedes** are about 1–3 inches long and can be as thick as 1 inch. The species found in Asia are larger and thicker.

- **These creatures are shiny black** or chestnut brown in color.

- **Pill millipedes** are nocturnal insects.

- **These millipedes** prefer to live in moist areas and are commonly found in evergreen forests. They are found in leaf litter and debris.

- **Although pill millipedes resemble woodlice**, woodlice have only seven pairs of legs, while pill millipedes have 17–19 pairs of legs.

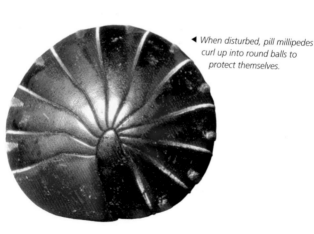

◄ When disturbed, pill millipedes curl up into round balls to protect themselves.

▲ *Like many other invertebrates, the pill millipede has a tough exoskeleton that provides both protection from predators and defense against drying out. Millipede exoskeletons are usually reinforced with calcium carbonate.*

- **Herbivorous**, pill millipedes feed on herbs, rotting leaves, moss, and organic matter.

- **Female pill** millipedes are longer than the males.

- **Females lay their eggs** in holes and fill the holes with tiny pellets of mud.

- **The young emerge** after two weeks looking like the adults, although they have fewer segments than adult millipedes.

Giant millipede

Giant millipedes belong to the order Spirostreptida. They are quite large and measure up to 12 inches in length. They have a segmented body and are either dark brown with golden lines on each segment or red with black lines. Each segment bears two pairs of legs.

These creatures are found in temperate and tropical regions of the world, including deserts, caves, forests, and along shorelines. Giant millipedes are abundant in tropical and arid coastal forests of eastern Africa.

Active at night, giant millipedes live in the soil or among plants. They feed on dead or decaying plant matter and have a life span of seven to ten years.

Giant millipedes have poor vision. Their antennae help them to find dead organic matter and decide whether it is decayed enough to eat.

The exoskeleton is made of calcium. Occasionally, the centipede chews on stones or pebbles.

To defend themselves, giant millipedes curl up into a spiral. This helps the millipede to protect its head and soft underside.

DID YOU KNOW?

Young giant millipedes are abandoned after hatching. They grow and mature slowly over three to ten years.

▶ The giant African millipede can grow up to 11 inches long.

🐦 **Giant millipedes** hide from predators beneath stones and plants. They can also produce a foul and irritating chemical from hidden glands.

🐦 **The male attracts** a female by moving along beside her and making her aware of the rhythmic pulses in his legs.

🐦 **A female makes** a small nest of compressed soil just below ground level. A few weeks after mating, she lays hundreds of eggs in this nest.

🐦 **The eggs** are covered with a tough, resistant coating to protect them from predators and adverse environmental conditions. The female guards the eggs until they hatch.

Index

Entries in **bold** refer to main subject entries; entries in *italics* refer to illustrations.